AIR CONDITIONING SYSTEM DESIGN

AIR CONDITIONING SYSTEM DESIGN

ROGER LEGG

Retired, previously senior lecturer at London
South Bank University

Butterworth-Heinemann
An imprint of Elsevier

Library of Congress Cataloging-in-Publication Data
A catalog record for this book is available from the Library of Congress

British Library Cataloguing-in-Publication Data
A catalogue record for this book is available from the British Library

ISBN: 978-0-08-101123-2

For information on all Butterworth-Heinemann publications
visit our website at https://www.elsevier.com/books-and-journals

www.elsevier.com • www.bookaid.org

Publisher: Matthew Deans
Acquisition Editor: Brian Guerin
Editorial Project Manager: Edward Payne
Production Project Manager: Anusha Sambamoorthy
Cover Designer: Mark Rogers

Typeset by SPi Global, India

DEDICATION

To staff and students,
past and present,
of the
'National College'.

The general antiphlogistic remedies are ... free admission of pure cool air.
John Alikin, 'Elements of Surgery', 1779

... the dreadful consequences which have been experienced from breathing air in situations either altogether confined or ill ventilated ... if others are in the same apartment, the breath from each person passes from one to another, and it is frequently in this way that diseases are communicated.
The Marquis de Chabannes, 1818

The very first rule of nursing ... is this: to keep the air he breathes as pure as the external air, without chilling him.
Florence Nightingale, 1863

CONTENTS

FOREWORD

Air conditioning is no longer regarded as the luxury that it once was, and there is now an increasing demand for applications ranging through domestic, commercial, industrial, and transport and for specialized installations such as hospitals, research facilities, data centres, and clean rooms. The engineering systems in modern buildings and installations make a significant contribution to the overall building performance in terms of energy use. Systems need to be increasingly sophisticated in their design, installation, operation, control, and maintenance at a time when there is increasing pressure for greater energy efficiency.

This has led to a demand for more qualified engineers and other professionals involved in building design. All those involved need to understand the underlying principles of the topics covered in this volume.

The book, which is a complete revision of Roger's previous work published by Batsford in 1991, contains new chapters on unitary systems and chilled beams. It provides a good technical foundation of building service engineering and covers significant proportions of the syllabus requirements of academic courses in this discipline. The theoretical coverage is backed with relevant worked examples and the use of data from the latest editions of CIBSE and ASHRAE publications, which should make this text appeal to students and practising professionals in both Europe and North America.

The author is well qualified in this discipline having taught the subject for more than 30 years at the Institute of Environmental Engineering (formerly the National College for Heating, Ventilation, Refrigeration and Fan Engineering, South Bank University, London). In addition, he has used contributions from key specialists to support specific areas; these included Associate Prof. Risto Kosonen, Prof. Tim Dwyer, Mr. Terry Welch, Prof. Ron James, Prof. John Missenden, and Mr. Stan Marchant.

Prof. Michael J. Farrell
London 2017
(Retired, previously principle lecturer at London South Bank University and head of the Institute of Environmental Engineering)

ACKNOWLEDGMENTS

I am indebted to my ex-colleagues at South Bank University for much practical help, encouragement, and advice in the writing of this book. In particular, I am most grateful to Mr. Terry Welch for VRV systems and the discussion in Chapter 7; to Prof. Ron James and Terry Welch for Chapter 9, on refrigeration and heat-pump systems; to Stan Marchant for the text on cooling towers in Chapter 10; to Prof. Tim Dwyer who contributed the overview of control systems in Chapter 17; and to Prof. John Missenden who provided the text for control valves in Chapter 17. My thanks are also due to Prof. Risto Kosonen of Aalto University, Sweden, for writing Chapter 8. My son Mark gave me a great deal of help with word processing. Lastly, my thanks are due to Brian Guerin, Edward Payne, and other members of Elsevier for their dedication in bringing this book to its completion.

BROMLEY 2017 RCL

The author and publishers thank the following for permission to use certain material from books and articles and to use illustrations as a basis for figures in this volume:

Tables 1.4 and 14.1 and Figs 1.16, 4.5, 7.2, 10.11, and 13.1 from the *CIBSE Guide* by permission of the Chartered Institute of Building Services Engineers.

Fig. 1.4 courtesy of the FISCHER company.

Fig. 3.2 (redrawn) by permission of McGraw Hill Book Co.

Table 4.4 warm temperatures in the United Kingdom, CIBSE Guide A.

Fig. 6.16 VAV Redrawn from Fig. 3.27 of the C1BSE Guide B, by permission of the Chartered Institute of Building Services Engineers.

Fig. 7.7 based on illustrations, courtesy of Trox Brothers Ltd.

Fig. 9.4 courtesy of ICI Chemicals and Polymers Ltd.

Fig. 10.6 drawing of jacketed steam humidifier based on Armstrong via website.

Plates 11.3 and 10.8 supplied by Thermal Technology Ltd.

Fig. 11.4 by permission of Fläkt Woods Limited.

Figs 12.2Bb and 11.4 supplied by Vokes Ltd.

Fig. 12.4 courtesy of Flaxt Woods—the United Kingdom.

Fig. 13.7 Moody chart from D S Miller Internal Flow Systems, Second edition, 1990, BHRA, Cranfield, the United Kingdom, with permission (note that the chart has some additional information that has been removed).

Figs 13.7, 14.5, 14.7, 14.11, and 14.13 (based on figures in Internal Flow Systems (Second Edition) 1990, BHRA, Cranfield, the United Kingdom) by permission of DS Miller.

Fig. 16.8 hooded vane anemometer, courtesy of Inlec the United Kingdom Ltd.

Fig. 16.9A courtesy of Holmes Valves Ltd.

Figs 16.11 and 16.13 courtesy Crane Fluid Systems.

Figs 16.9B, 16.11, and 16.15 courtesy of Crane Ltd.

Fig. 19.1 by permission of the Building Services Research and Information Association.

CHAPTER 1

Properties of Humid Air

Air is the working fluid for air conditioning systems. It is therefore important for the engineer to have a thorough understanding of the properties of air, before going on to consider the processes that occur when air passes through the various plant items that make up systems. The word *psychrometry* is often used for the science that investigates the properties of humid air, and the chart that shows these properties graphically is known as the *psychrometric* chart.

In this chapter, the various air properties are defined, and the appropriate equations are given. In deriving the equations, it is usual to consider the air as consisting of two gases, dry air and water vapour. Even though one of these is strictly a vapour, both are considered to obey the ideal gas laws. Lastly, the tables and chart, from which numerical values of the air properties are obtained for practical calculations, are described and illustrated.

ATMOSPHERIC PRESSURE

At any point in the earth's atmosphere, there exists a pressure due to the mass of air above that point—the atmospheric pressure. Standard atmospheric pressure at sea level is 1013.25 mbar (usually approximated to 1013 mbar), but due to changes in weather conditions, there are variations from this standard pressure. For example, among the minimum and maximum values recorded in London are 948.7 mbar (in 1821) and 1048.1 mbar (in 1825), respectively; those recorded for North America are 892 mbar (Long Key, Florida, in 1935) and 1074 mbar (Yukon Territory, Canada, in 1989) [1].

Atmospheric pressure varies with height above sea level, and for altitudes at which mankind lives, the rate of decrease (lapse rate) for a standard atmosphere may be taken as a *reduction* of 0.13 mbar per meter of height *above* sea level and an *increase* of 0.13 mbar per meter of depth *below* sea level.

Air Conditioning System Design
http://dx.doi.org/10.1016/B978-0-08-101123-2.00001-7

Example 1.1

Determine the standard atmospheric pressure for Nairobi, which is at an altitude of 1820 m above sea level.

Solution

Standard sea-level atmospheric pressure	1013
Lapse rate $= -1820 \times 0.13$	-237
Standard atmospheric pressure for Nairobi	776 mbar

Atmospheric pressure may be measured by using a number of instruments. In the laboratory, it is usual to use a Fortin barometer, while for site work an aneroid barometer is the most usual instrument. For continuous recording, a barograph is used.

DRY AIR AND WATER VAPOUR

Dry air consists of a number of gases but mainly of oxygen and nitrogen. It is necessary to know the molecular mass of the dry air, and this is calculated from the proportion each individual gas makes in the mixture. Table 1.1 gives this data, together with the calculation.

The sum of the molecular mass fractions is 28.97 and this is the value taken as the mean molecular mass of dry air.

Water vapour is said to be *associated* with the dry air. Its molecular mass is obtained from the masses of its chemical composition H_2O, i.e.,

$$M_{H_2O} = (2 \times 1.01) + (1 \times 16) = 18.02$$

Table 1.1 Determination of molecular mass of dry air

Gas	Proportion by volume (%) (1)	Molecular mass (%) (2)	Molecular mass fraction (%) (1) × (2)
Nitrogen, N_2	78.03	28.02	21.86
Oxygen, O_2	20.99	32.00	6.72
Carbon dioxide, CO_2	0.03	44.00	0.01
Hydrogen, N_2	0.01	2.02	0.00
Argon, Ar	0.92	39.91	0.38
Molecular mass fraction			28.97

VAPOUR PRESSURE

Saturated Vapour Pressure

Consider the vessel shown in Fig. 1.1. The contents are at temperature 1°C, and the atmosphere above the water contains water vapour that exerts a pressure known as *saturated vapour pressure* (SVP). When heat is applied to the vessel, more water evaporates, and as the temperature rises, the SVP increases. Eventually, with heat still being supplied, the water will boil, and this happens when the SVP is equal to atmospheric pressure. The variation of saturated vapour pressure against temperature is shown in Fig. 1.2.

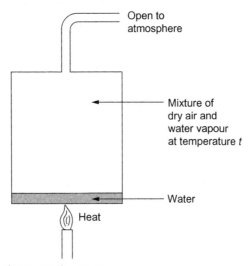

Fig. 1.1 Vessel with saturated vapour.

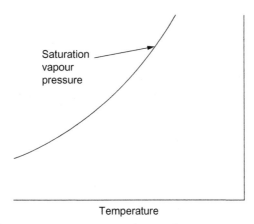

Fig. 1.2 Saturation vapour pressure versus temperature.

Table 1.2 Saturation vapour pressures

Dry-bulb temperature (°C)	SVP (mbar)	Dry-bulb temperature (°C)	SVP (mbar)
0	6.103	20	23.37
2	7.055	25	31.66
4	8.129	30	42.42
6	9.346	35	56.22
8	10.72	40	73.75
10	12.29	50	123.4
12	14.02	60	199.2
14	15.98	70	311.6
16	18.17	80	473.6
18	20.63	100	1013.2

Values of SVP have been determined by experiment and published in the form of steam tables, selected values of which are given in Table 1.2.

There is no simple relationship between temperature and SVP. The following equations are the relevant curve fits published by the National Engineering Laboratory [2]:

For water above 0°C,

$$\log_{10} P_{\text{ssw}} = 28.59 - 8.2\log_{10} T + 0.00248T - 3142/T$$

where P_{ssw} is the SVP in bar, over water at absolute T (K).

For ice below 0°C:

$$\log_{10} = 10.538 - 2664/T$$

where P_{ssi} is the SVP in bar, over ice at absolute temperature T (K).

These equations are suitable for use in computer programs in which air property values are required; they are not used in this text.

Superheated Vapour

If all the water in the vessel shown in Fig. 1.2 evaporates before boiling point has been reached and heat continues to be applied, the water vapour becomes superheated with the vapour pressure remaining constant. Therefore, on Fig. 1.2, the superheated vapour is in the region to the right-hand side of the SVP curve. Air conditioning engineers will normally be interested only in the variations in vapour pressure in the temperature range from −20°C to 60°C.

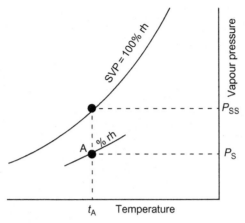

Fig. 1.3 Definition of relative humidity.

RELATIVE HUMIDITY

Definition—Relative humidity is the percentage ratio of the vapour pressure of water vapour in the air to the saturated vapour pressure at the same temperature.

From the definition, relative humidity of air at condition **A** in Fig. 1.3 is therefore given by:

$$\varphi = \frac{P_s}{P_{ss}} \times 100 \tag{1.1}$$

100%rh corresponds to the saturated vapour pressure P_{ss}.

Example 1.2
Air at 20°C has a vapour pressure of 13 mbar. Determine the relative humidity.

Solution
From Table 1.2, the SVP at 20°C is 23.37 mbar. Using Eq. 1.1:

$$\varphi = \frac{P_s}{P_{ss}} \times 100 = \frac{13}{23.37} \times 100 = 55.6\%$$

From the discussion on vapour pressure, it should be noted that atmospheric pressure only determines the boiling point of water; it has no effect on saturated vapour pressure or vapour pressure, and therefore, atmospheric pressure has no effect on relative humidity.

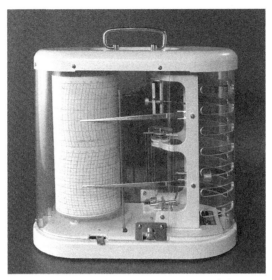

Fig. 1.4 Thermohygrograph. *(Reproduced with permission of the FISCHER company.)*

Relative humidity can be measured directly by a number of instruments, in particular with a thermohygrograph as illustrated in Fig. 1.4. However, for more accurate measurements and for calibrating other humidity measuring devices, it is more usual to measure it indirectly by using dry- and wet-bulb temperature measurements. These can then be referred to tables of humid air properties or to a psychrometric chart, to determine the relative humidity.

IDEAL GAS LAWS

The ideal gas laws are used to derive a number of humid air properties. The errors in the numerical values of the air properties due to departures from the ideal laws are very small. For a discussion on this point, see Jones [3].

Dalton's Law of Partial Pressures

Dalton's law of partial pressures states that the pressure of a mixture of gases is equal to the sum of the partial pressure that each individual gas would exert by itself at the same volume and temperature.

$$P_t = P_1 + P_2 + P_3 \tag{1.2}$$

where P_t is total pressure of the mixture of gases and P_1, P_2, P_3 are the partial pressures of the individual gases.

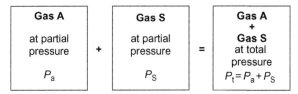

Fig. 1.5 Dalton's law of partial pressures.

Dalton's law is illustrated in Fig. 1.5; two gases **A** and **S** at pressures P_a and P_s, which individually occupy the same volume, are combined in one of the vessels to give the total pressure, P_t.

Example 1.3

If the atmospheric pressure is 1013 mbar and the water vapour pressure is 40 mbar, determine the partial pressure of the dry air.

Solution

Total air pressure (atmospheric)	1013
Partial pressure water vapour	−40
Partial pressure of dry air	973 mbar

General Gas Law

Boyle's law states that, at constant temperature, the product of the pressure p and volume V of a gas remains constant, i.e.,

$$pV = \text{constant}$$

Charles's law states that the volume of a gas V is proportional to its absolute temperature T, the pressure remaining constant, i.e.,

$$V/T = \text{constant}$$

Boyle's and *Charles's* laws combine to give the general gas law:

$$pV = mRT \tag{1.3}$$

Note that in the absolute temperature $T = (273 + t_a)$, the air dry-bulb temperature t_a is in degrees Celsius. Individual gas constants are calculated from the universal gas constant R_o and the molecular mass M of the gas, i.e.,

$$R = R_o/M \tag{1.4}$$

The value of R_o is 8314.66 J/kmol K, and the molecular mass is expressed in kg/kmol.

Example 1.4

Determine the gas constants for dry air and water vapour.

Solution

As determined previously, the molecular masses for dry air and water vapour are 28.97 and 18.02, respectively. The gas constants are therefore calculated as follows:
Using Eq. (1.4):
for dry air,

$$R_a = 8314.66/28.97 = 287 \ J/kgK$$

for water vapour,

$$R_s = 8314.66/18.02 = 461 \ J/kgK$$

DENSITY OF AIR

The density of air can be calculated using the general gas equation, and this is illustrated by the following example.

Example 1.5

Determine the density of air with a temperature 20°C (normal temperature) and at an atmospheric pressure of 1013 mbar (standard pressure).

Solution

The air density is given by:

$$\rho = \frac{m_a}{V} = \frac{P_{at}}{R_a T} = \frac{1013 \times 100}{287 \times (273 + 20)} = 1.205 \ kg/m^3$$

The standard value of air density is usually taken as 1.20 kg/m^3. Again, consideration of the general gas law shows how the air density can be corrected for atmospheric pressure and temperatures that differ from those on which the standard is based, i.e.,

$$\rho = 1.2 \times \frac{P_{at}(273 + 20)}{1013(273 + t_a)}$$

$$\rho = 0.347 \times \frac{P_{at}}{(273 + t_a)} \qquad (1.5)$$

Example 1.6

Determine the density of air, which has a temperature of 30°C and an atmospheric pressure of 980 mbar.

Solution

Using Eq. (1.5), the air density is calculated as: given by

$$\rho = 0.347 \times \frac{P_{at}}{(273 + t_a)} = 0.347 \frac{980}{(273 + 30)} = 1.122 \, kg/m^3$$

MOISTURE CONTENT

Definition—The moisture content of humid air is the mass of water vapour present in 1 kg of dry air.

This air property is variously referred to as *humidity ratio, specific humidity or is calculated as absolute humidity.*

It is important to recognize at this point in the discussion on humid air that some of its properties are based on 1 kg of *dry air*, unlike the properties of most other fluid mixtures, which are based on 1 kg of the mixture.

The derivation is as follows:

Using the general gas law, Eq. (1.3),

for dry air:

$$p_a V_a = m_a R_a T_a$$

for water vapour:

$$p_s V_s = m_s R_s T_s$$

For the mixture, $V_a = V_s$ and $T_a = T_s$. Therefore, from the definition of moisture content given above:

$$g = \frac{m_s}{m_a} = \frac{R_a p_s}{R_s p_a} = \frac{287 p_s}{461 p_a}$$

From Dalton's law of partial pressures, Eq. (1.2):

$$p_a = p_{at} - p_s$$

$$g = 0.622 \frac{p_s}{(p_{at} - p_s)} \qquad (1.6)$$

Example 1.7

Determine the moisture content for air at a temperature of 20°C and a vapour pressure of 13 mbar when the atmospheric pressure is 1013 mbar.

Solution

Using Eq. (1.6):

$$g = 0.622 \frac{p_s}{(p_{at} - p_s)} = 0.622 \frac{13}{1013 - 13}$$

$$= 0.00809 \,\mathrm{kg/kg_{da}}$$

SATURATION MOISTURE CONTENT

If the vapour pressure p_s in Eq. (1.6) is at SVP p_{ss}, then the moisture content becomes the saturation moisture content. In the same way that saturated vapour pressure varies with temperature, saturation moisture content also varies with temperature. This is illustrated graphically in Fig. 1.6, the resulting curve being a prominent feature of the psychrometric chart. Some typical values of saturation moisture contents are given in Table 1.3.

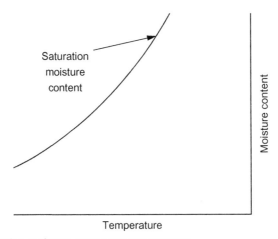

Fig. 1.6 Saturation moisture content vs temperature.

Table 1.3 Saturation moisture contents

Dry-bulb temperature (°C)	Moisture content (kg/kg$_{da}$)	Dry-bulb temperature (°C)	Moisture content (kg/kg$_{da}$)
0	0.00379	20	0.0147
2	0.00438	25	0.0202
4	0.00505	30	0.0273
6	0.00582	35	0.0367
8	0.00668	40	0.0491
10	0.00766	45	0.0653
12	0.00876	50	0.0868
14	0.01001	60	0.153
16	0.0114		
18	0.01300		

PERCENTAGE SATURATION

Definition—Percentage saturation is the percentage ratio of the moisture content in the air to the moisture content at saturation at the same temperature.

The percentage saturation of air at condition **A** in Fig. 1.7 is therefore given by:

$$\mu = \frac{g_s}{g_{ss}} \times 100 \tag{1.7}$$

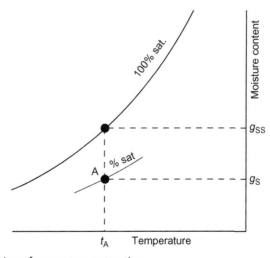

Fig. 1.7 Definition of percentage saturation.

Example 1.8

Calculate the percentage saturation for the air condition described in Example 1.7.

Solution

From the solution of Example 1.7, for the air condition specified, the moisture content is 0.00809 kg/kg$_{da}$. From Table 1.3, the saturation moisture content at 20% is 0.0147 kg/kg$_{da}$.

Using Eq. (1.7):

$$\mu = \frac{g_s}{g_{ss}} \times 100 = \frac{0.00809}{0.0147} \times 100$$

$$= 0.55$$

Lines of constant percentage saturation appear on a psychrometric chart as shown in Fig. 1.8. For practical purposes, values of percentage saturation are interchangeable with those of relative humidity. Percentage saturation is slightly dependent upon atmospheric pressure and accounts for the small numerical differences that exist between these two air properties and that will be noted in the tables of air properties.

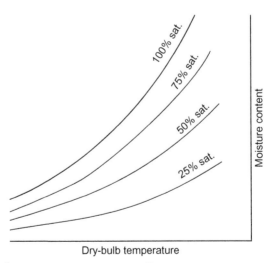

Fig. 1.8 Lines of constant percentage saturation.

SPECIFIC VOLUME

Definition—Specific volume is the volume of air containing 1 kg of dry air plus the associated moisture content. The derivation is as follows:

Using the general gas law for dry air, Eq. (1.3):

$$p_a V_a = m_a R_a T_a$$

From the definition given above:

$$v = \frac{V_a}{m_a} = \frac{R_a T_a}{p_a}$$

From Dalton's law of partial pressures:

$$p_a = p_{at} - p_s$$

$$v = \frac{(273 + t_a)}{(p_{at} - p_s)}$$

$$v = \frac{287(273 + t_a)}{(p_{at} - p_s)} \tag{1.8}$$

Example 1.9

Determine the specific volume for air at 20%, a vapour pressure of 14 mbar, and an atmospheric pressure of 1013 mbar.

Solution

Using Eq. (1.8):

$$v = \frac{287(273 + t_a)}{(p_{at} - p_s)} = \frac{287 \times (273 + 200)}{(1013 - 14) \times 100}$$
$$= 0.842 \, \text{m}^3/\text{kg}_{da}$$

Lines of constant specific volume are drawn on the psychrometric chart as shown in Fig. 1.9.

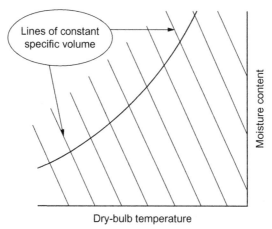

Fig. 1.9 Lines of constant specific volume.

Relationship Between Air Density and Specific Volume

Air density is defined as the mass of air per *unit* volume, whereas the specific volume is defined in terms of *unit* mass of dry air. Therefore, the relationship between the two is:

$$\rho = (1 + g)\, v \tag{1.9}$$

Though the difference between density and the reciprocal of specific volume is relatively small, the engineer should be aware of the difference compared with the true relationship, as given in Eq. (1.9), when making calculations in different areas of work. Thus, it is usual to use density when measuring airflow rates through pressure drop devices such as orifice plates and to use specific volume in air conditioning load calculations.

DRY-BULB AND WET-BULB TEMPERATURES

Dry- and wet-bulb temperatures, measured together, are among the most popular methods for determining the air condition, and from these measurements, other air properties may be derived. Dry- and wet-bulb temperatures can be measured using a variety of instruments, e.g., mercury-in-glass, thermocouple, and resistance thermometers.

Dry-bulb Temperature

Definition—The dry-bulb temperature of air is the temperature obtained with a thermometer, which is freely exposed to the air but which is shielded from radiation and free from moisture. The word *dry* is used to make a distinction from the wet-bulb.

Wet-bulb Temperature

Definition—The wet-bulb temperature of air is the temperature obtained with a thermometer whose bulb is covered by a muslin sleeve that is kept moist with distilled/clean water, freely exposed to the air, and free from radiation. The reading obtained is affected by air movement over the instrument. For this reason, there are two wet-bulb temperatures—*sling* and *screen*:

(1) The sling wet-bulb is obtained in a moving air stream, preferably above 2 m/s. This is usually measured with either a sling hygrometer (Fig. 1.10) or an Assman hygrometer. However, a sling reading may also be obtained if a wet-bulb thermometer is installed in a duct through which air is flowing at a reasonable velocity. The sling wet-bulb is considered to be more accurate than the screen wet-bulb temperature and for this reason is preferred by air conditioning engineers.

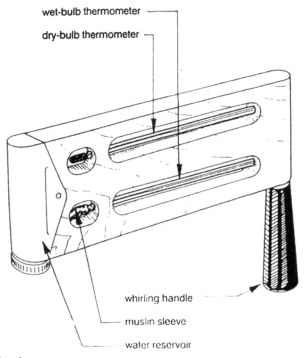

wet-bulb thermometer

dry-bulb thermometer

whirling handle

muslin sleeve

water reservoir

Fig. 1.10 Sling hygrometer.

(2) The screen wet-bulb is assumed to be in still air, usually installed in a Stevenson screen (from which this type of wet-bulb derives its name), as used by meteorologists.

The Psychrometric Equation

The psychrometric equation relates the dry- and wet-bulb temperatures with their corresponding vapour pressures and with the atmospheric pressure. To understand this relationship, consider the diagram of the wet-bulb thermometer in Fig. 1.11.

Moisture is being evaporated from the surface of the muslin sleeve into the surrounding air. For evaporation to take place, heat must be supplied, and this can only come from the ambient air in the form of sensible heat, with the temperature of the bulb lower than that of the surrounding air. At equilibrium, the latent heat loss due to moisture evaporation will equal the sensible heat gained. The air film at the surface of the muslin sleeve is considered to be at saturation moisture content g_{ss}'. (Note the $'$ to indicate that the moisture content is at the wet-bulb temperature.) The latent heat loss is proportional to the moisture content difference between this air film and the ambient air, i.e., $(g_{ss}' - g)$. The sensible heat gained is proportional to the temperature difference between the bulb and the ambient air $(t - t')$, i.e.,

$$B(g_{ss}' - g) = C(t - t') \tag{1.10}$$

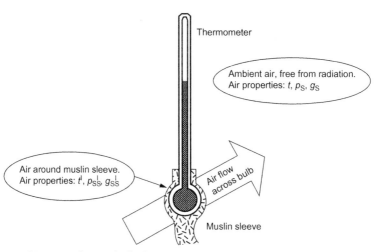

Fig. 1.11 Diagram of a wet-bulb thermometer.

where B and C are constants related to parameters of heat and mass transfer, e.g., surface area and latent heat of evaporation.

From Eq. (1.6):

$$g = 0.622 \frac{p_s}{(p_{at} - p_s)}$$

and

$$g = 0.622 \frac{p'_{ss}}{(p_{at} - p'_{ss})}$$

Since p_s and p_{ss}' are very small compared with p_{at}, these equations may be written as:

$$g = 0.622 \frac{p_s}{p_{at}} \quad \text{and} \quad g'_{ss} = 0.622 \frac{p'_{ss}}{p_{at}}$$

Substituting these expressions of moisture content in Eq. (1.10),

$$\frac{0.622B}{p_{at}} (p'_{ss} - p_s) = C(t - t')$$

By rearranging the terms and grouping the constants, the psychrometric equation is obtained:

$$p_s = p'_{ss} - p_{at}A(t - t') \tag{1.11}$$

where A is known as the *psychrometric constant*.

The numerical difference between the dry- and wet-bulb temperatures is known as the *wet-bulb depression*.

Since the rate of moisture evaporation depends on the speed of the air over the wet-bulb, the wet-bulb temperature will also depend on the air speed. However, the wet-bulb becomes independent of the air velocity above 2 m/s. The two wet-bulb temperatures described above—sling and screen—cater for this with different values for the constant A.

Wet-bulb temperatures are also affected by the air being either above or below freezing point, and again, different values of A are necessary to deal with these conditions. The psychrometric constants for a 4.8 mm bulb diameter are the following:

sling

$$A = 6.66 \times 10^{-4}\,\text{K} \quad \text{when} \quad t' > 0°C$$

$$A = 5.94 \times 10^{-4}\,\text{K} \quad \text{when} \quad t' < 0°C$$

screen
$$A = 7.99 \times 10^{-4}\,\text{K}^{-1} \quad \text{when} \quad t' > 0°\text{C}$$
$$A = 7.20 \times 10^{-4}\,\text{K}^{-1} \quad \text{when} \quad t' < 0°\text{C}$$

When working with the psychrometric equation, it is important to remember that the saturated vapour pressure p_{ss}' is taken at the wet-bulb temperature.

Example 1.10

Calculate the vapour pressure for air with the following conditions:

Dry-bulb temperature	22%
Wet-bulb temperature (sling)	14°C
Atmospheric pressure	1013 mbar

Solution

From Table 1.2, the SVP at 14°C $= 15.98$ mbar. Since the air is above 0°C and the wet-bulb is a sling reading, the psychrometric constant A is 6.66×10^{-4} K.
Using Eq. (1.11):

$$p_s = p_{ss}' - p_{at}A\,(t - t')$$
$$p_s = 15.98 - 1013 \times 6.66 \times 10^{-4}) \times (22 - 14)$$
$$= 11.93 \text{ mbar}$$

Lines of constant wet-bulb temperature are drawn on the psychrometric chart, as illustrated in Fig. 1.12.

DEW-POINT TEMPERATURE

Definition—The dew-point temperature is the temperature of saturated air that has the same vapour pressure as the air condition under consideration.

Referring to Fig. 1.13, when air at the original condition **A** is cooled at constant vapour pressure, i.e., at constant moisture content, the temperature of the air will eventually reach the saturation line, and at this point, water vapour will begin to condense. This temperature, a unique condition on the saturation line, is known as the dew-point temperature t_{dp}.

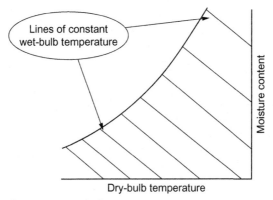

Fig. 1.12 Lines of constant wet-bulb temperature.

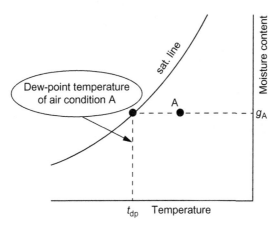

Fig. 1.13 Dew-point temperature.

Example 1.11

Air at a dry-bulb temperature of 40°C and having a moisture content of 0.202 kg/kg$_{da}$ is cooled at constant vapour pressure. At what temperature will dew begin to form?

Solution

Cooling air at constant vapour pressure is the same as cooling it at constant moisture content. Referring to Table 1.3, the saturation moisture content of 0.0202 kg/kg$_{da}$ occurs at 25°C. Therefore, the dew-point temperature of the given air condition is 25°C.

There are commercially available instruments that measure dew-point temperature directly. However, it is more usual to obtain its value by referring to tables of properties of humid air or to psychrometric chart, using measurements of other air properties such as dry-bulb and wet-bulb temperatures.

SPECIFIC ENTHALPY

Definition—The specific enthalpy of humid air is a calculated property combining the sensible and latent heat of 1 kg of dry air plus its associated water vapour, relative to a datum at 0°C.

The equation for specific enthalpy is formulated as follows:

Consider 1 kg of dry air and the associated moisture content 'g' at dry-bulb temperature 't.' The sensible heat h_1 of 1 kg of dry air, relative to the datum 0°C, is given by:

$$h_1 = 1 \times c_{pa}(t - 0)$$

where:

$$c_{pa} = \text{specific heat of dry air}$$

$$= 1.005 \text{ kJ/kgK}$$

$$h_1 = 1.005\,t \qquad\qquad (1.12)$$

Similarly, the sensible heat of the moisture content h_2 is given by:

$$h_2 = g c_{ps}(t - 0)$$

where:

$$c_{pa} = \text{specific heat of water vapour}$$

$$= 1.89 \text{ kJ/kgK}$$

$$h_2 = 1.89 g t \qquad\qquad (1.13)$$

The water vapour is considered to have evaporated at 0°C, and therefore the latent heat of the moisture content h_3 is given by:

$$h_3 = g h_{fg}$$

where h_{fg} = the latent heat of evaporation at $0°C$ = 2501 kJ/kg

$$\therefore h_3 = 2501 g \tag{1.14}$$

The specific enthalpy of humid air, h, is obtained from the sum of Eqs (1.12)–(1.14):

$$h = h_1 + h_2 + h_3$$

$$h = 1.005\, t + g\,(1.89t + 2501) \tag{1.15}$$

Example 1.12

Determine the specific enthalpy of air at a dry-bulb temperature of $20°C$ and moisture content of 0.008 kg/kg$_{da}$.

Solution

Using Eq. (1.15):

$$h = 1.005t + g\,(1.89t + 2501)$$
$$= 1.005 \times 20 + 0.008\,(1.89 \times 20 + 2501)$$
$$= 40.4\,\text{kJ/kg}_{da}$$

Lines of constant enthalpy do not always appear on a psychrometric chart. For example, on the CIBSE chart (Fig. 1.15), to obtain the specific enthalpy of an air condition, a straight edge is used to join the corresponding enthalpy marks above the 100% saturation line with those on either the bottom or the right-hand side of the chart.

Humid Specific Heat

Eq. (1.15) for specific enthalpy can be rearranged as follows:

$$h = (1.005 + 1.89g)\,t + 2501g$$

$$= c_{pas}\, t + 2501g$$

where:

$$c_{pas} = 1.005 + 1.89g \tag{1.16}$$

The term c_{pas} is known as the *humid specific heat*.

Example 1.13

Calculate the humid specific heat for the air condition in Example 1.12.

Solution

Using Eq. (1.16), the humid specific is given by

$$c_{pas} = 1.005 + 1.89g = 1.005 + 1.89 \times 0.008$$
$$= 1.02\,kJ/kg_{da}\,K$$

The variations in the moisture content found in atmospheric air are such that for practical purposes an average value of $1.02\,kg/kg_{da}\,K$ may be used for c_{pas} in sensible heat load calculations.

ADIABATIC SATURATION TEMPERATURE

An adiabatic process is one in which no external heat enters or leaves the system under consideration. In Fig. 1.14, air is flowing through a duct in the bottom of which is an open-water tank.

The plant casing is considered to be perfectly insulated so that no heat flows into the duct from the surroundings or vice versa. Air enters the duct at dry-bulb temperature t_1 and moisture content g_1, and as it passes down the duct, moisture will be evaporated so that at the end of the duct the air will have a moisture content g_2. For water to evaporate, heat must be supplied, and since this is an adiabatic process, this can come only from the air itself. Therefore, the latent heat gained by the air must equal the sensible heat loss by the air. In other words, there must be a drop in air dry-bulb temperature to compensate for the increase in moisture content. If the air leaves at dry-bulb temperature t_2, then for each kilogram of dry air:

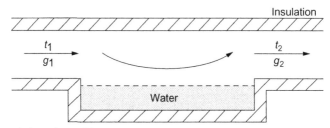

Fig. 1.14 Adiabatic humidification process.

latent heat gained = sensible heat loss

$$\therefore (g_2 - g_1)h_{fg} = c_{pas}(t_1 - t_2) \tag{1.17}$$

If the tank is infinitely long, then the air at the end of the process will beat 100% saturation and the temperature at which this occurs is known as *adiabatic saturation temperature, t^**, and the corresponding moisture content is g_{ss}^*. By substituting t^* for t_2 and g_{ss}^* for g_2 in Eq. (1.17) and rearranging,

$$t^* = t - (g_{ss}^* - g)h_{fg}/c_{pas} \tag{1.18}$$

To determine values of the adiabatic saturation temperature t^*, an iterative solution of Eq. (1.18) is required.

TABLES OF PROPERTIES

The properties of humid air at the standard atmospheric pressure of 1013.25 mbar are given in tables published by CIBSE [4]. As an example of these tables, the properties for a dry-bulb temperature of 20°C are given in Table 1.4. The special condition that is 100% saturation—with the relative humidity also at 100%—should be noted. At this point, the dry-bulb, wet-bulb, dew-point, and adiabatic saturation temperatures are equal, and the vapour pressure is the saturated vapour pressure.

When specifying an air condition, it is usual to give the dry-bulb temperature and one other property. From these two values, the other properties can be obtained. If the value of any property is not uniquely specified in the table, linear interpolations between adjacent conditions are justified.

THE PSYCHROMETRIC CHART

The psychrometric chart is a most useful design tool for air conditioning engineers. A typical chart is shown in Fig. 1.15 [4]. The air properties on the chart are:
- dry-bulb temperature,
- sling wet-bulb temperature,
- moisture content,
- specific enthalpy,
- specific volume,
- percentage saturation.

Table 1.4 Properties of humid air at 20°C dry-bulb temperature

								Wet-bulb temperature	
Percentage saturation, μ (%)	Relative humidity, ϕ (%)	Moisture content, $g/(\text{g kg}^{-1})$	Specific enthalpy, $h/(\text{kJ kg}^{-1})$	Specific volume, $v/(\text{m}^3\,\text{kg}^{-1})$	Vapour pressure, p_v (kPa)	Dew-point temperature, θd (°C)	Adiabatic saturation temperature, θ^* (°C)	Screen, θ'_{sc} (°C)	Sling, θ'_{st} (°C)
100	100.00	14.75	57.55	0.8497	2.337	20.0	20.0	20.0	20.0
96	96.09	14.16	56.05	0.8489	2.246	19.4	19.6	19.6	19.6
92	92.17	13.57	54.56	0.8481	2.154	18.7	19.1	19.2	19.1
88	88.25	12.98	53.06	0.8473	2.062	18.0	18.7	18.8	18.7
84	84.31	12.39	51.56	0.8466	1.970	17.3	18.2	18.3	18.2
80	80.37	11.80	50.06	0.8458	1.878	16.5	17.7	17.9	17.7
76	76.43	11.21	48.57	0.8450	1.786	15.7	17.2	17.5	17.3
72	72.47	10.62	47.07	0.8442	1.694	14.9	16.7	17.0	16.8
70	70.49	10.33	46.32	0.8438	1.647	14.5	16.5	16.8	16.5
68	68.51	10.03	45.57	0.8434	1.601	14.0	16.2	16.5	16.3
66	66.53	9.736	44.82	0.8431	1.555	13.6	16.0	16.3	16.0
64	64.54	9.441	44.08	0.8427	1.508	13.1	15.7	16.1	15.8
62	62.55	9.146	43.33	0.8423	1.462	12.6	15.5	15.8	15.5
60	60.56	8.851	42.58	0.8419	1.415	12.1	15.2	15.6	15.3
58	58.57	8.556	41.83	0.8415	1.369	11.6	14.9	15.4	15.0
56	56.58	8.260	41.08	0.8411	1.322	11.1	14.7	15.1	14.7
54	54.59	7.966	40.33	0.8407	1.276	10.6	14.4	14.9	14.5
52	52.59	7.670	39.58	0.8403	1.229	10.0	14.1	14.6	14.2
50	50.59	7.376	38.84	0.8399	1.182	9.4	13.9	14.4	13.9

Value of stated parameter per kg dry air

48	48.59	7.080	38.09	0.8395	1.136	8.8	13.6	14.1	13.7
46	46.59	6.785	37.34	0.8391	1.089	8.2	13.3	13.9	13.4
44	44.58	6.490	36.59	0.8388	1.042	7.6	13.0	13.6	13.1
42	42.58	6.195	35.84	0.8384	0.9945	6.9	12.7	13.4	12.8
40	40.57	5.900	35.09	0.8380	0.9480	6.2	12.4	13.1	12.5
38	38.56	5.605	34.34	0.8376	0.9011	5.5	12.1	12.8	12.2
36	36.55	5.310	33.60	0.8372	0.8541	4.7	11.8	12.6	12.0
34	34.53	5.015	32.85	0.8368	0.8070	3.9	11.5	12.3	11.7
32	32.52	4.720	32.10	0.8364	0.7600	3.0	11.2	12.0	11.4
30	30.50	4.425	31.35	0.8360	0.7127	2.1	10.9	11.8	11.1
28	28.48	4.130	30.60	0.8356	0.6656	1.2	10.6	11.5	10.7
24	24.43	3.540	29.10	0.8348	0.5710	-0.8	10.0	10.9	10.1
20	20.38	2.950	27.61	0.8341	0.4763	-3.0	9.3	10.4	9.5
16	16.32	2.360	26.11	0.8333	0.3814	-5.6	8.6	9.8	8.8
12	12.25	1.770	24.61	0.8325	0.2863	-8.9	8.0	9.2	8.2
8	8.17	1.180	23.11	0.8317	0.1910	-13.4	7.3	8.6	7.5
4	4.09	0.590	21.62	0.8309	0.0956	-20.8	6.5	8.0	6.8
0	0.00	0.000	20.11	0.8301	0.0000	—	5.8	7.3	6.1

Reproduced from Section C1 of the *CIBSE Guide*, with the permission from the Chartered Institute of Building Services Engineers

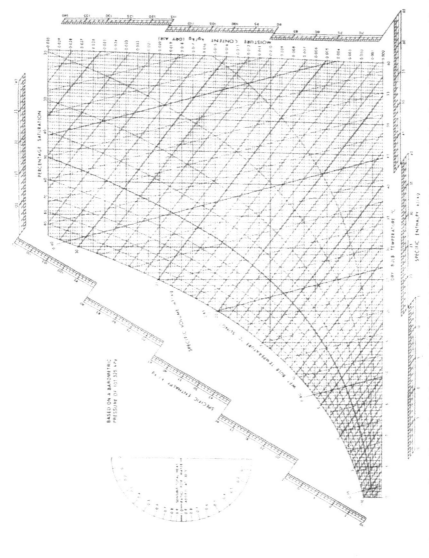

Fig. 1.15 Psychrometric chart. (*Reproduced from Section C1 of the CIBSE Guide, with the permission from the Chartered Institute of Building Services Engineers.*)

When working with the chart, the following points should be noted:

- Sling wet-bulb is used in preference to screen wet-bulb temperature as it is considered to be the more consistent of the two measurements.
- Vapour pressure is not given. This property is rarely required by the air conditioning engineer.
- Dew-point temperature can be obtained for a given air condition as previously described.
- The lines of constant dry-bulb temperature are not at right angles to the lines of constant moisture content. This is because the chart is based on enthalpy and moisture content and dry-bulb temperature is added subsequently, determined from the enthalpy equation. (See Example 2.1 in the following chapter.)

SYMBOLS

A	psychrometric constant
B, C	constants
c_{pas}	specific heat of *humid* air
g	moisture content
g_{ss}	saturation moisture content
g_{ss}^*	adiabatic saturation moisture content
h	specific enthalpy of *humid* air
h_{fg}	latent heat of evaporation of water
M	molecular mass
m	mass
p	pressure
p_{at}	atmospheric pressure
p_{ss}	saturation vapour pressure
p_{ss}'	saturation vapour pressure at wet-bulb temperature
R	particular gas constant
R_o	universal gas constant
T	absolute temperature
t	dry-bulb temperature
t'	wet-bulb temperature, sling
t_{sc}'	wet-bulb temperature, screen
t^*	adiabatic saturation temperature
t_{dp}	dew-point temperature
V	volume
v	specific volume of humid air
μ	percentage saturation
ρ	air density
φ	% relative humidity

SUBSCRIPTS

a   air, dry air
s   water vapour (steam)

ABBREVIATIONS

SVP   saturation vapour pressure
%rh   percentage relative humidity
%sat   percentage saturation

REFERENCES

[1] P. Eden, Independent Radio News; personal communication, 1989 (pre-adjusted to equivalent sea level pressures).
[2] Mayhew, Rogers, Steam Tables, fifth ed., Blackwell Publishing, 2014.
[3] W.P. Jones, Air Conditioning Engineering, fifth ed., Edward Arnold, London, 2001.
[4] CIBSE, Guide Book C: Reference Data, 2007.

CHAPTER 2

Air Conditioning Processes

Air conditioning plant items can be thought of as the *building blocks* from which systems are designed and constructed. It is necessary to understand the psychrometric processes that can be achieved with each *block* before dealing with the complete *system*. These processes can be shown most easily on a psychrometric chart or as a psychrometric sketch as in the diagrams that follow. The air conditions are given as letters, and these correspond with those on a diagram of the equipment itself. The process line on the chart is usually shown as a straight line, even though the actual conditions of the air as it passes *through* the plant item might, to some extent, deviate from that line. Generally, the air conditioning systems engineer is interested only in the state of the air as it *enters* and leaves the item of plant.

MIXING OF TWO AIR STREAMS

Airstreams at different conditions are often mixed within an air conditioning system, the most usual case being that of air from outdoors (via the fresh-air intake grille) mixing with air returned from the air conditioned space.

Fig. 2.1 shows two airstreams, **A** and **B**, mixing to produce condition **M**. It is assumed that the mixing process is adiabatic, i.e., there is no leakage of air into or out of the ductwork and no miscellaneous heat gains or losses. Because of the way, the psychrometric chart has been constructed and from the laws of conservation of mass and energy, the mixing process can be drawn as the straight line **AMB** on the chart. If 'x' is the proportion of airstream **A** in the total air mass flow rate leaving the system, then the air properties of the mixed-air condition are determined as follows:

Specific enthalpy:

$$h_M = x h_A + (1 - x) h_B \qquad (2.1)$$

Moisture content:

$$g_M = x g_A + (1 - x) g_B \qquad (2.2)$$

Dry-bulb temperature (approximately):

$$t_M = x t_A + (1 - x) t_B \qquad (2.3)$$

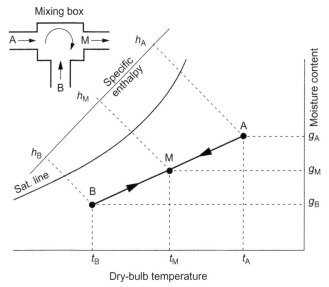

Fig. 2.1 Mixing of two airstreams.

Example 2.1

In an air conditioning plant, 0.5 kg/s of outdoor air mixes with 1.5 kg/s of recirculated air. Determine the specific enthalpy, moisture content, and dry-bulb temperature of the mixed airstream for the following air conditions:

Airstream	Dry-bulb temperature (°C)	Moisture content (kg/kg$_{da}$)
Recirculated air (**A**)	22	0.010
Outdoor air (**B**)	4	0.002

Solution

Air mass flow rate for the mixture $= 0.5 + 1.5 = 2.0$ kg/s
%age of outdoor air to total air, $x = 0.5/2.0 = 0.25$
Refer to Fig. 2.2.
The specific enthalpies of the airstreams **A** and **B** are obtained from tables of air properties (or less precisely from a psychrometric chart):

$$h_A = 47.54 \, \text{kJ/kg}_{da} \quad \text{and} \quad h_B = 9.04 \, \text{kJ/kg}_{da}$$

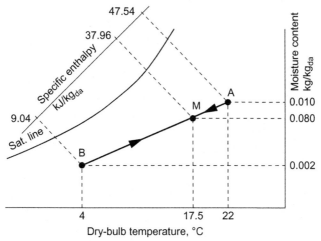

Fig. 2.2 Air mixing—Example 2.1.

The mixed-air enthalpy is obtained from Eq. (2.1):

$$h_M = xh_A(1-x)h_B$$
$$= 0.25 \times 9.04 + (1-0.25)47.54 = 37.92\,kJ/kg_{da}$$

The mixed-air moisture content is obtained from Eq. (2.2):

$$g_M = xg_A + (1-x)g_B$$
$$= 0.25 \times 0.002 + (1-0.25)0.01 = 0.008\,kg/kg_{da}$$

The mixed-air temperature is obtained from Eq. (2.3), approximately:

$$t_M = xt_A + (1-x)t_B$$
$$= 0.25 \times 4 + (1-0.25)22 = 17.5°C$$

The value of the dry-bulb temperature determined from Eq. (2.3) is sufficiently accurate for practical air conditioning calculations. If a precise value is required, the temperature of the mixed airstream t_M should be calculated using the enthalpy Eq. (1.15), with values of specific enthalpy and moisture content determined from Eqs (2.1) and (2.2), respectively.

Example 2.2

Using the data from Example 2.1, calculate a precise value for the mixed-air temperature.

Solution

Using Eq. (1.15):

$$h_M = (1.005 + 1.89g_M)t_M + 2501g_M$$
$$37.92 = (1.005 + 1.89 \times 0.008)t_M + 2501 \times 0.008$$
$$\therefore t_M = (37.96 - 20.01)/1.02 = 17.6°C$$

This value of the mixed-air temperature agrees closely with the approximate value determined in Example 2.1.

SENSIBLE HEATING COILS

A sensible heating process, occurring at constant moisture content, is one in which the dry-bulb temperature of the air is increased when the air passes over a hot, dry surface. The heater might be a pipe coil using hot water or steam or electrical resistance elements or one of a number of the alternative air-to-air heat recovery units described in Chapter 10. Heaters are required in air conditioning systems for frost protection, as preheaters for humidifiers and as afterheaters to maintain space temperatures.

In Fig. 2.3, air passes through such a heater, the air dry-bulb temperature rising from condition A to condition B, the moisture content remaining

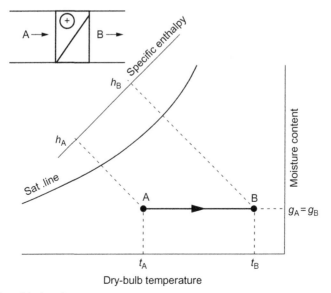

Fig. 2.3 Sensible heating.

constant. The specific enthalpy of the air will also increase. The load on the heater battery is then given by either one of the following two equations:

$$Q_h = \dot{m}_a c_{pas}(t_B - t_A) \tag{2.4}$$

$$Q_h = \dot{m}_a(h_B - h_A) \tag{2.5}$$

The choice of equation will depend on the function of the heater within the system. Thus, Eq. (2.4) is usually appropriate for a heater that deals with room or zone sensible loads that are related to the dry-bulb temperature, whereas Eq. (2.5) is preferred for a preheater used in conjunction with an adiabatic humidifier when the heating load is related to both sensible and latent heat exchange in the plant. However, both equations will give the same load.

Example 2.3
An air flow rate of 1.5 kg/s passes through a heater battery, the dry-bulb temperature rising from 10°C to 24°C. Calculate the load on the heater battery.

Solution
Refer to Fig. 2.4. Using air temperatures and Eq. (2.4):

$$Q_h = \dot{m}_a c_{pas}(t_B - t_A)$$
$$= 1.5 \times 1.02(24 - 10) = 21.4\,kW$$

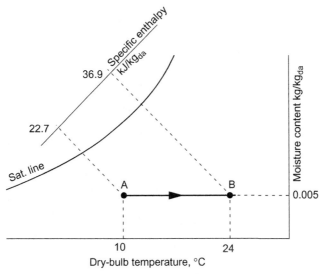

Fig. 2.4 Sensible beating—Example 2.3.

Using Eq. (2.5) and assuming a moisture content of 0.006 kg/kg, the enthalpies are obtained from either tables or chart:

$$Q_h = \dot{m}_a(h_B - h_A)$$
$$= 1.5(36.9 - 22.7) = 21.3\,kW$$

The small difference between the two loads is due to rounding errors; the difference is not significant for practical calculations.

COOLING COILS

Coolers are required in air conditioning systems to reduce the supply-air temperature and moisture content, particularly in summer when there are heat gains to the rooms, to provide for both room temperature and humidity control. An air-cooling coil uses either a heat-transfer medium such as chilled water or the direct expansion of refrigerant into a pipe coil. The cooling of air may be sensible only, or it may be accompanied by dehumidification giving latent and a sensible cooling load for the coil.

Sensible Cooling at Constant Moisture Content

With sensible cooling, the air temperature is reduced, and for this to occur at constant moisture content, all parts of the coil (air-side) surface temperature must be above the dew-point temperature of the entering airstream.

In Fig. 2.5, air passes through a cooler, the air dry-bulb temperature falling from condition A to condition B, the moisture content remaining

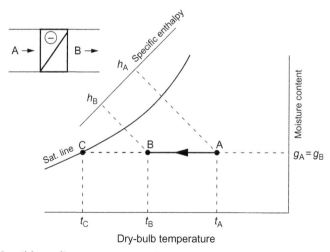

Fig. 2.5 Sensible cooling at constant moisture content.

constant. The load on the cooling coil is then given by either of the following equations:

$$Q_c = \dot{m}_a c_{pas}(h_A - h_B) \tag{2.6}$$

$$Q_c = \dot{m}_a(h_A - h_B) \tag{2.7}$$

Eq. (2.6) would usually be used to determine the load, since no latent heat exchange is involved. But both equations will produce the same answer.

Sensible Cooling With Dehumidification

In Fig. 2.6, the air passes through a cooler, both the air dry-bulb temperature and the moisture content failing from condition A to condition B. The heat in the water that has been condensed from the airstream will normally be very small relative to the total cooling load and therefore can be ignored in the calculations. The load on the cooling coil is then given by Eq. (2.7). For the coil to dehumidify, the coil (air-side) surface temperature must be below the dew-point temperature of the entering airstream. The average coil temperature at air condition **C**, on the saturation line, is known as the *apparatus dew-point temperature* (ADP), and the line **ABC** is drawn as a straight line on the chart.

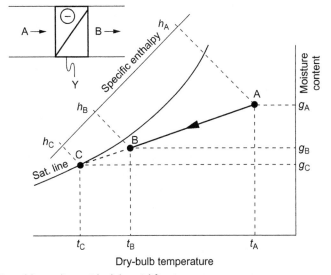

Fig. 2.6 Sensible cooling with dehumidification.

Example 2.4

An air flow rate of 2.4 kg/s passes through a cooling coil. Determine the load on the coil for the following air conditions:

Airstream	Dry-bulb temperature (°C)	Moisture content (kg/kg$_{da}$)
On coil (A)	24	0.0100
Off coil (B)	12	0.0075

Solution

Refer to Fig. 2.7. From the psychrometric chart, the specific enthalpies are as follows:

On coil $\qquad\qquad\qquad\qquad h_A = 49.5 \text{ kJ/kg}_{da}$

Off coil $\qquad\qquad\qquad\qquad h_B = 31.0 \text{ kJ/kg}_{da}$

Using Eq. (2.6),

$$Q_c = \dot{m}_a(h_A - h_B)$$

$$= 2.4(49.5 - 31) = 44.4 \text{kW}$$

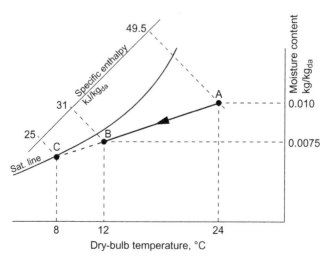

Fig. 2.7 Sensible cooling with dehumidification—Example 2.4.

Cooling Coil Contact Factor

An important property of a coil is its ability to dehumidify. Referring to Fig. 2.6, the minimum moisture content for a coil that is 100% efficient at dehumidifying the air would be the saturation moisture content g_C at the apparatus dew-point. The (decimal) efficiency for dehumidification, known as the contact factor of the coil, is therefore defined by any one of the following equations:

By moisture content differences:

$$\beta = \frac{g_A - g_B}{g_A - g_C} \qquad (2.8)$$

By specific enthalpy differences:

$$\beta = \frac{h_A - h_B}{h_A - h_C} \qquad (2.9)$$

By dry-bulb temperature differences (approximately):

$$\beta = \frac{t_A - t_B}{t_A - t_C} \qquad (2.10)$$

Example 2.5

Using the data in Example 2.4, determine the contact factor of the coil.

Solution

The process line has to be drawn on a psychrometric chart, as in Fig. 2.7. The continuation of the line **AB** cuts the saturation line at **C**. For the conditions given, $t_C = 8°C$ at which condition $h_C = 25$ kJ/kg.
Using Eq. (2.9),

$$\beta = \frac{h_A - h_B}{h_A - h_C}$$
$$= \frac{49.5 - 31}{49.5 - 25} = 0.76$$

The actual value of the contact factor will depend on the design of the coil and the air flow rate. The most important parameters are the following:

- Number of rows of pipe coils;
- Design of the heat-transfer surface;
- Air velocity across the face of the coil;
- Drainage of condensate.

The plant diagram in Fig. 2.6 shows a drain, trap, and tundish to deal with the water that has been condensed from the airstream. This is a practical requirement, the trap preventing air from being blown out of, or drawn into, the airstream. The break between the trap and the tundish ensures that the air conditioning system is not contaminated through the drain; it also provides the plant operator with a convenient means to observe whether or not the coil is dehumidifying.

HUMIDIFIERS

Humidifiers are used to increase the moisture content of the air within air conditioning systems, and these may be classified into three groups:
1. Adiabatic with recirculation of the spray water;
2. Adiabatic with no recirculation of the spray water;
3. Isothermal.

The psychrometric operating characteristics of these humidifiers are given here. For a more detailed description of these humidifiers, see Chapter 10.

Adiabatic Humidifiers: Recirculation of Spray Water

This group of humidifiers includes the spray washer, the capillary washer, and the sprayed–cooling coil. Fig. 2.8 shows a diagram of a typical unit. Water is pumped from a tank mounted at the bottom of the humidifying

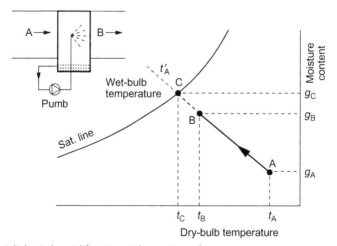

Fig. 2.8 Adiabatic humidification with an air washer.

chamber, and the water that is not evaporated into the airstream falls back into the tank to be recirculated by the pump through the pipework to the nozzles. The theoretical process line follows the adiabatic saturation temperature, though in practice this is usually drawn following the wet-bulb temperature of the entering air condition. However, a close approximation, which assists the solution of a number of air conditioning calculations, is to consider the process occurring at constant specific enthalpy.

The make-up water to be supplied to the humidifier tank is given by:

$$\dot{m}_w = \dot{m}_a(g_B - g_A) \qquad (2.11)$$

The humidifying efficiency is usually expressed as a contact factor, which is defined by either of the following equations:

By moisture content differences:

$$\beta = \frac{g_B - g_A}{g_C - g_A} \qquad (2.12)$$

By dry-bulb temperature differences (approximately):

$$\beta = \frac{t_B - t_A}{t_C - t_A} \qquad (2.13)$$

The contact factor depends on the nozzle design, the number and arrangement of spray nozzles, and the water pressure at the nozzles. In the capillary type, water is sprayed on to cells packed with a suitable material that allows close contact between air and water. In the sprayed coil type, the wetted air-side surface of the cooling coil provides the majority of the surface from which water is evaporated into the air.

Example 2.6
Air at 25°C dry-bulb temperature and 15°C wet-bulb temperature enters a spray water humidifier with a contact factor of 0.7. Determine the moisture content of the air leaving the humidifier.

Solution
Refer to Fig. 2.9. From the psychrometric chart the moisture content of the air entering the humidifier is obtained:

$$g_A = 0.0065 \, kg/kg_{da}$$

Assuming the humidifying process occurs at constant wet-bulb temperature, the ADP is also at 15°C at which condition:

$$g_C = 0.0107 \, kg/kg_{da}$$

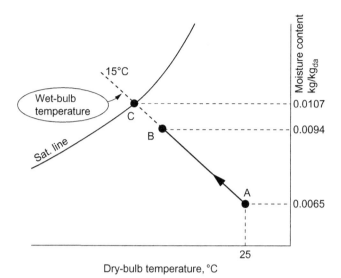

Fig. 2.9 Adiabatic humidification—Example 2.6.

Using Eq. (2.12):

$$\beta = \frac{g_B - g_A}{g_C - g_A}$$

$$0.7 = \frac{g_B - 0.0065}{0.0107 - 0.0065}$$

$$g_B = 0.0094 \, \text{kg/kg}_{da}$$

Spray Humidifiers as Heat Exchangers—Nonadiabatic Processes

Spray-type humidifiers with water recirculation can be designed with heating and cooling equipment to achieve heating, cooling, and dehumidification processes, in addition to the adiabatic humidification process previously described.

The pumped water heated by a calorifier is shown in Fig. 2.10. As the heat supply to the calorifier increases, the temperature of the spray water rises producing a corresponding rise in the ADP, above the adiabatic saturation temperature (point **C**) of the air entering the process. The process line will be one of a set, typically shown by **AB₁**, **AB₂**, and **AB₃**, depending on the amount of heat supplied to the spray water. With the AB₁ process, the increase in moisture content is accompanied by some sensible cooling.

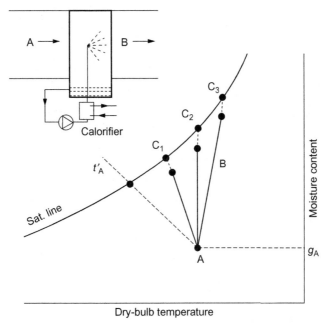

Fig. 2.10 Air washer with heating of spray water.

The **AB₂** process is the special case of humidification occurring at constant dry-bulb temperature, i.e., isothermally. In the **AB₃** process, humidification is occurring with sensible heating. In each case, the heating load on the calorifier is given by:

$$Q_h = \dot{m}_a(h_B - h_A) \tag{2.14}$$

The humidifying contact factor is given by Eq. (2.12). It may also be defined in terms of specific enthalpy differences, i.e.,

$$\beta = \frac{h_B - h_A}{h_A - h_C} \tag{2.15}$$

Example 2.7

An air flow rate of 2.4 kg/s at a condition of 9°C dry-bulb temperature and 3°C wet-bulb temperature enters a spray water humidifier. If 40 kW of heat is provided by a calorifier to the spray water, and given that the contact factor is 0.8, determine the moisture content of the air leaving the humidifier.

Solution

Refer to Fig. 2.11. From the chart, enthalpy and moisture content of air entering process is given as:

$$h_A = 14.5 \, \text{kJ/kg}_{da}$$
$$g_A = 0.00215 \, \text{kg/kg}_{da}$$

Using Eq. (2.14):

$$Q_h = \dot{m}_a(h_B - h_A)$$
$$40 = 2.4(h_B - 14.5)$$
$$h_B = 31.2 \, \text{kJ/kg}_{da}$$

Using Eq. (2.15):

$$\beta = \frac{h_B - h_A}{h_A - h_C}$$

$$0.7 = \frac{31.2 - 14.5}{h_C - 14.5}$$

$$h_C = 38.4 \, \text{kJ/kg}_{da}$$

The process line is now drawn on the chart, joining condition **A** to condition **C** with an enthalpy of 38.4 kJ/kg$_{da}$ at which condition the moisture content $g_C = 0.0097$ kg/kg$_{da}$. Using Eq. (2.12):

$$\beta = \frac{g_B - g_A}{g_C - g_A}$$

$$0.7 = \frac{g_B - 0.00215}{0.0097 - 0.00215}$$

$$g_B = 0.00744 \, \text{kg/kg}_{da}$$

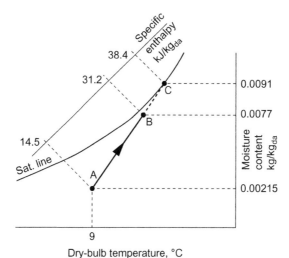

Fig. 2.11 Air washer with heating of spray water—Example 2.7.

Sprayed-Cooling Coil

Consider the case of the sprayed-cooling coil shown in Fig. 2.12. With the coil receiving no coolant, the humidifying process will be adiabatic, with the ADP at condition **C**. With the coil receiving coolant, the ADP will be depressed below giving a set of process lines, typically **AB₄**, **AB₅**, and **AB₆**. With the **AB₄** process C, humidification is occurring with sensible cooling, the **AB₅** process is the special case of sensible cooling at constant moisture content, and the **AB₆** process is one of sensible cooling with dehumidification, the same process previously described for the dehumidifying cooling coil. The cooling load will be given by Eq. (2.7) and the contact factor for cooling by Eqs (2.8)–(2.10).

At first sight, it may seem strange that the processes **AB₅** and **AB₆** occur when water is being sprayed into the airstream. They do occur because the coil air-side surface temperature is at, or below, the dew-point temperature of the entering air.

The processes described for the sprayed-cooling coil can also be obtained with the air washer by introducing chilling to the spray water. However, this once popular method of including a whole range of psychrometric processes within one piece of equipment is rarely used today because of the probability of scaling of heat-transfer surfaces and problems associated with the hydraulic circuits, e.g., maintaining water levels in a water-chilling refrigeration plant above the level of the humidifier tank.

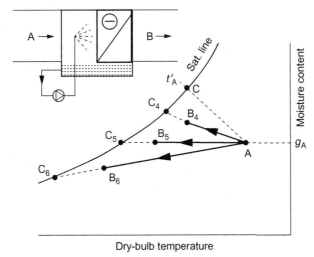

Dry-bulb temperature

Fig. 2.12 Sprayed-cooling coil processes.

Adiabatic Humidifiers: No Recirculation of Spray Water

These humidifiers include the spinning disk and direct water injection through a nozzle, the psychrometric process being shown in Fig. 2.13.

The amount of water supplied is for humidification purposes only, none is recirculated, and any water not evaporated in the process is drained from the bottom of the unit. The efficiency of humidification, expressed as a contact factor and calculated using Eqs (2.12) and (2.13), depends on the fineness of the water droplets produced, either by the operation of the spinning disk or by the pressurized water in the nozzle. Only a small amount of water is supplied relative to the mass of air passing through the unit. Therefore, heating or chilling the water supply will not bring about a significant departure from an adiabatic process, and the practical process may be taken as following the wet-bulb temperature of the entering air condition.

Not all the water supplied to the humidifier is evaporated into the airstream, and the water supplied to the unit must include for the overall humidifying efficiency that is defined by

$$\eta_h = \frac{\dot{m}_w}{\dot{m}_a(g_B - g_A)} \times 100 \qquad (2.16)$$

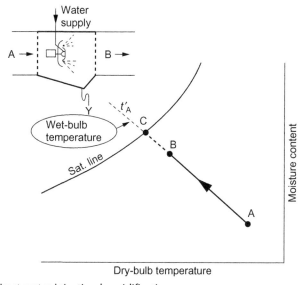

Fig. 2.13 Direct water injection humidification.

Example 2.8

An air mass flow rate of 1.4 kg/s enters a spinning disk humidifier and leaves with a moisture content of 0.009 kg/kg$_{da}$. If the overall humidifying efficiency is 60%, determine the water supply rate to the humidifier. The moisture content of the entering air is 0.005 kg/kg$_{da}$.

Solution

Using Eq. (2.16):

$$\eta_h = \frac{\dot{m}_w}{\dot{m}_a(g_B - g_A)} \times 100$$

$$60 = \frac{\dot{m}_w}{1.4(0.009 - 0.005)} \times 100$$

$$\dot{m}_w = 0.00336\,\text{kg/s} = 3.36\,\text{L/s}$$

Steam Humidifiers

At present, steam humidifiers are the most popular method of increasing the moisture content of the air supply in air conditioning systems. There are two basic types:

1. Direct steam injection;
2. Pan.

Direct Steam Injection

Steam that is injected directly into the airstream is supplied either from a central boiler or through a local steam generator unit installed as part of the air conditioning system. With this type of humidifier, all the latent heat for evaporation is added outside the airstream, and the water vapour supplied to the air increases the enthalpy of the air. Since it occurs at near-constant temperature, the process is usually referred to as an isothermal process.

Referring to Fig. 2.14, the load on the humidifier is given by:

$$Q_s = \dot{m}_a(h_B - h_A) \tag{2.17}$$

The steam supplied by the humidifier is given by:

$$\dot{m}_s = \dot{m}_a(g_B - g_A) \tag{2.18}$$

The small rise in dry-bulb temperature from a steam humidifier is due to the sensible heating effect of the steam. This temperature rise is calculated by

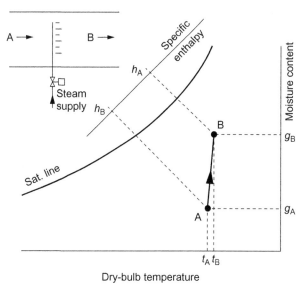

Fig. 2.14 Steam humidification.

equating the sensible heat gained by the air to the sensible heat loss by the steam, i.e.,

$$\dot{m}_a c_{pas}(t_B - t_A) = \dot{m}_a(g_B - g_A)c_{ps}(t_S - t_B)$$

Example 2.9

An air supply rate of 1.4 kg/s of air at 15°C and with a moisture content of 0.004 kg/kg$_{da}$ enters a steam humidifier. Calculate the load on the humidifier and the steam supplied, if the air leaves with a moisture content of 0.008 kg/kg$_{da}$.

Solution

Referring to Fig. 2.15 and assuming the process occurs at constant temperature. The enthalpies are obtained from the psychrometric chart:

$$h_A = 25 \, kJ/kg_{da} \quad and \quad h_A = 35.5 \, kJ/kg_{da}$$

Using Eq. (2.14):

$$Q_h = \dot{m}_a(h_B - h_A)$$
$$= 1.4(35.5 - 25) = 14.7 \, kW$$

The quantity of steam supplied is obtained using Eq. (2.18):

$$\dot{m}_s = \dot{m}_a(g_B - g_A)$$
$$= 1.4(0.008 - 0.004) = 0.0056 \, kg/s$$

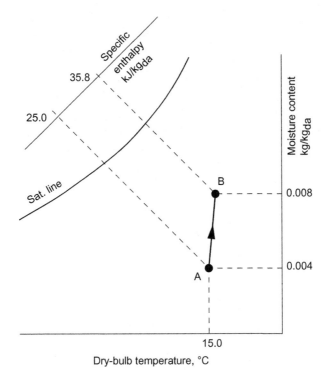

Fig. 2.15 Steam humidification—Example 2.9.

The specific heats are $c_{pas} = 1.025$, $c_{pas} = 1.89$, and the steam temperature is 100°C. Since $(t_s - t_B)$ will not be significantly different from $(t_s - t_A)$, the temperature rise in the air is given by:

$$(t_B - t_A) = 1.84(g_B - g_A)(100 - t_A) \qquad (2.19)$$

Example 2.10
Determine the temperature rise of the air for the steam humidifier in Example 2.9.

Solution
Using Eq. (2.19),

$$(t_B - t_A) = 1.84(g_B - g_A)(100 - t_A)$$

$$= 1.84(0.008 - 0.004)(100 - 15) = 0.63 \text{K}$$

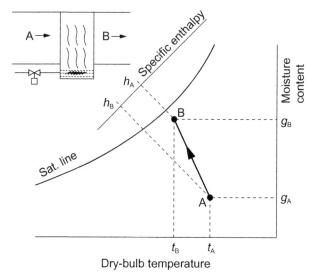

Fig. 2.16 Steam pan humidification.

In practice, the process line for a steam humidifier may be drawn at constant temperature, as assumed in Example 2.9.

Pan Steam Humidifier

A pan-type steam humidifier consists of a water tank mounted at the bottom of the air duct, together with a heating element. When heat is supplied, vapour is evaporated from the surface of the water to produce the humidifying effect. Some evaporative cooling also occurs, caused by the air flowing over the water surface, and because of this, there will be a corresponding drop in air dry-bulb temperature. This will be equivalent to a sensible-to-total heat ratio of approximately 0.2 (Fig. 2.16). (Sensible-to-total heat ratios on the psychrometric chart are described in the following section, relative to room process lines.)

The heat load on the humidifier and the steam supply are obtained from Eqs (2.17) and (2.18), respectively.

Adiabatic Dehumidification

In locations where there is high humidity (as in the tropics), it may be appropriate to reduce the moisture content of the incoming air by using an adiabatic dehumidifier (rather than dehumidify with a cooling coil as described above). The process is shown in Fig. 2.17; incoming air **A** passes through the dehumidifier, and moisture is removed. Since the process is adiabatic, there

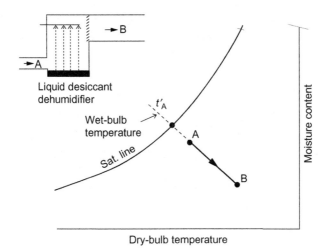

Fig. 2.17 Adiabatic dehumidification.

will be an increase in sensible heat content, and the temperature increases to state point **B**. As with adiabatic humidification, the theoretical process line follows the adiabatic saturation temperature (near-constant specific enthalpy or the wet-bulb temperature).

Dehumidifiers use either solid or liquid desiccants. Solid desiccants are used in a wheel similar to a heat recovery thermal wheel (as described in Chapter 11). The wheel is packed with an absorbent silica gel; as the wheel rotates, incoming air **A** passes through the dry gel where water vapour is absorbed; exhaust air passing in the opposite direction 'regenerates' the gel that is then discharged to atmosphere. About 60% of the wheel is used for the dehumidification process.

This process is similar with a liquid-desiccant system. The high humidity airstream **A** passes through the salt solution where moisture is absorbed; again the desiccant is regenerated to be reused in, as it were, a continuous cycle. Eliminator plates will be necessary to prevent carry-over in the supply airstream.

If an adiabatic dehumidifier is used in a comfort air conditioning system, then the temperature of **B** will need to be reduced to an appropriate supply-air temperature with a sensible cooling coil as shown in Fig. 2.5.

AIR CONDITIONED SPACE: ROOM PROCESS LINES

To gain an understanding of the psychrometric processes that occur in an air conditioned space, it is useful to consider the sensible loads to be produced by either a cooler or a heater and the latent loads by either a humidifier or a dehumidifier.

Taking the case of a room receiving both a sensible and latent heat gain in Fig. 2.18, the room is maintained at air condition **R** and is supplied with air at condition **S**. The heat balances are as follows:

$$q_s = \dot{m}_a c_{pas}(t_R - t_S) \tag{2.20}$$

$$q_1 = \dot{m}_a h_{fg}(g_R - g_S) \tag{2.21}$$

The ratio of the room sensible-to-latent heat gains is obtained by dividing Eq. (2.20) by Eq. (2.21), i.e.,

$$\frac{q_s}{q_l} = \frac{c_{pas}(t_R - t_S)}{h_{fg}(g_R - g_S)}$$

With average values of $c_{pas} = 1.025\,\text{kJ/kgK}$ and $h_{fg} = 2450\,\text{kJ/kg}$, the following relationship is obtained:

$$\frac{q_s}{q_l} = 0.000418\frac{(t_R - t_S)}{(g_R - g_S)} \tag{2.22}$$

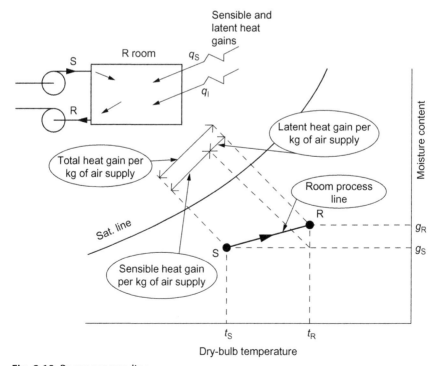

Fig. 2.18 Room process line.

Eq. (2.22) shows that, for a given room sensible-to-latent heat ratio, the temperature to moisture content ratio of the supply-air is independent of the supply-air mass flow rate.

Room Ratio Lines

The room process line is usually known as the *room ratio line* (RRL), and any supply condition on that line will satisfy the room condition with an appropriate air supply rates. When designing, it will be usual to draw the RRL on the psychrometric chart before the supply condition is determined, and the most usual method of doing this is to make use of the pair of quadrants printed on the chart. Referring to Fig. 2.19, the procedure is as follows:

1. Calculate the sensible-to-total heat gain ratio 'y':

$$y = \frac{q_s}{q_s + q_l} \tag{2.23}$$

2. Enter the value of 'y' on the appropriate quadrant, in this case, the bottom quadrant. Measure the angle of 'y' from the horizontal line (1.0), and this will give the slope of the RRL. Draw a line at this angle on the chart, passing through the required room condition, **R**.

In calculating the value of 'y', all room loads are taken as positive values, irrespective of whether the load is a gain or a loss. The differences between loads are dealt with by the two quadrants forming a semicircle, and the

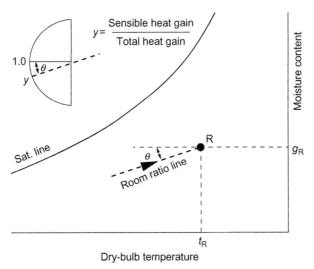

Fig. 2.19 Plotting the room ratio line.

choice of which one to use will depend on the orientation of the RRI, on the chart. Depending on whether the room loads are gains or losses, there are four possible positions of the RRL, i.e.,

- sensible heat gain and latent heat gain—bottom quadrant;
- sensible heat loss and latent beat gain—top quadrant;
- sensible heat gain and latent heat loss—bottom quadrant;
- sensible heat loss and latent heat loss—top quadrant.

SYMBOLS

c_{ps} specific heat of steam
c_{pas} specific heat of humid air
g moisture content
h specific enthalpy of humid air
h_{fg} latent heat of evaporation of water
$\dot{m}_a$ mass flow rate of dry air
$\dot{m}_s$ mass flow rate of steam
$\dot{m}_w$ mass flow rate of water
Q_c cooling load
Q_h heating load
Q_s humidifier load
q_s sensible heat gain to air conditioned space
q_1 latent heat gain to air conditioned space
t dry-bulb temperature
x fraction of outdoor air
γ ratio of room sensible-to-total heat gain
β contact factor
η_h humidifying efficiency

SUBSCRIPTS

A, B   relate to specific air conditions
C   air condition on the saturation line
M   mixed-air condition
R   room air condition

ABBREVIATION

ADP   apparatus dew-point temperature
RRL   room ratio line

CHAPTER 3

Indoor Design Conditions

When a building is to be air conditioned, it will be necessary for the designer to decide the internal space conditions that should be maintained throughout the year when the building is occupied. Many systems are required to provide conditions that meet the thermal comfort conditions for the occupants; other systems provide conditions suitable for the efficient operation of machines and processes, the storage of food and artefacts. However, it will rarely be the case that air conditions have to be maintained at a constant level; variations are usually permitted about an optimum level. In this chapter, various aspects of these topics are examined, leading to a choice of appropriate indoor design conditions.

THERMAL COMFORT

In normal health, a man or a woman has an internal body temperature of about 37°C, and this temperature has to be supported for healthy living. Departures of a few degrees from normal body temperature are usually a sign of ill health, and even a danger to life itself; heat stroke and hypothermia are well known, if relatively rare, examples of high and low body temperatures.

The body generates a certain amount of heat due to the oxidation of food, and this has to be dissipated if the body temperature is not to rise. Conversely, if too much heat is lost to the surroundings, the body temperature will fall. The amount of heat produced will depend on the level of physical activity or rate of work. At rest, the body produces about 100 W of heat and during hard work about 500 W. It is the body's physiological mechanisms that regulate the rate of heat production together with the rate of heat loss, arranging the balance that maintains near-constant body temperature. When the body temperature falls, more heat is generated by imperceptible tensing of the muscles, a further fall by the onset of shivering. A rise in body temperature is countered by increased perspiration.

The nude body can only cope with a small range of external conditions in maintaining its body temperature, and clothing is therefore used as insulation. The amount of clothing will affect the rate of heat loss, and this in turn will affect the feeling of *warmth*.

Air Conditioning System Design
http://dx.doi.org/10.1016/B978-0-08-101123-2.00003-0

The feeling of warmth depends on a balance between the rate of heat production and the rate of heat loss, which in turn depends on the environmental conditions. Generally, the body loses heat through *convection, radiation,* and *evaporation,* but it may also gain heat by convection and radiation when the surrounding air and surface temperatures are higher than the body's surface temperature. Evaporation heat loss consists of insensible perspiration from the skin together with the water vapour expired from the lungs.

When physical activity is low, in a normal indoor environment, the proportion of heat loss through these modes of heat transfer is of the order of 45%, 30%, and 25%, respectively. Relative to the surface conditions of the body are the following:

- Convection heat loss (or gain) depends on air dry-bulb temperature and air velocity.
- Radiation heat loss/gain depends on the temperatures of the surrounding room surfaces, including the surfaces of heat-producing equipment within the room. The average temperature of all these surfaces is usually expressed as the *mean radiant temperature* (MRT).
- Evaporation heat loss through insensible perspiration depends on the air vapour pressure and the air velocity.

The four variables of the physical environment that affect heat loss from the body are therefore:

- air dry-bulb temperature,
- air vapour pressure (or relatively humidity),
- air velocity,
- mean radiant temperature.

For any individual, the sensation of thermal comfort is a complex subjective reaction to an environment, depending on a number of *personal factors,* such as age, sex, and state of health. However, for a group of people, there are only two personal factors that have a significant correlation with comfort, these being:

- the amount of physical activity (rate of work),
- the amount of clothing worn.

INDICES OF THERMAL COMFORT

The air dry-bulb temperature in itself is most often a satisfactory index by which thermal comfort can be judged, but where the other variables have a significant effect on the rate of heat exchange, they should be taken into account. The CIBSE index combining air temperature and MRT as a single

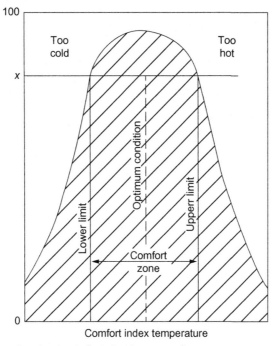

Fig. 3.1 Diagram showing basis for selecting a comfort zone.

numerical value, is the *operative temperature* [1]. ASHRAE uses the term *effective temperature* [2]. The scale of each of these indices is a temperature against which the subjective assessment of thermal comfort can be measured.

COMFORT REQUIREMENTS

The literature on thermal comfort is now considerable. The first major work on the subject in the United Kingdom was carried out by Bedford [3]. In these original studies, Bedford established optimum conditions and comfort zones[1] for workers engaged in light manual or sedentary work. The comfort conditions were based on the comfort votes of a large group of people. At the optimum condition, there were a small percentage of those who felt some thermal discomfort, e.g., there might be a feeling of being slightly cool or slightly warm. As the comfort temperature deviates from the optimum condition, the level of dissatisfaction increases. This is illustrated in Fig. 3.1.

---

[1] A comfort zone is an range of (comfort) temperatures in which there is a risk of a certain number of people experiencing discomfort.

Table 3.1 Comfort zones of factory workers engaged in light manual or sedentary work with range of moisture contents

	Winter			Summer	
Thermal index	Lower limit (°C)	Optimum condition (°C)	Upper limit (°C)	Optimum condition (°C)	Upper limit (°C)
Air temperature	15.5	18.5	22.0	19.5	24.0
Globe temperature	15.5	18.5	23.5	20.6	24.0
Equivalent temperature	14.8	16.7	21.0	19.0	23.0

Twenty years or so after Bedford's research, Hickish determined the requirements for comfort for a similar group of workers in summer. The work by Hickish indicated that the zones established by Bedford were still appropriate, but with the optimum summer condition, 1–2°K higher compared with winter [4]. This difference, similar to the findings of researchers in the United States, was mainly attributable to the lighter clothing worn in summer compared with winter. Though it has been suggested that the standards of thermal comfort have risen since these original studies, it is probable that the comfort zones still apply.

The temperatures applicable to thermal comfort, obtained by Bedford and Hickish, are given in Table 3.1. The comfort zones are based on a satisfaction level (comfort vote) of 80% of the occupants. A lower limit was not available from the survey of summer conditions. It was considered that the upper limit is determined by the onset of sweating.

THE COMFORT EQUATION

The most commonly accepted work on thermal comfort is that by Fanger who derived the *comfort equation* [5]. This equation is based on the heat balance for the human body when in thermal equilibrium with the environment, correlated with the two personal factors of *activity level* and *clothing*, the four environmental factors and subjective observations of the state of comfort.

The internal heat produced by the body during a particular activity is expressed in units known as the *met*, one met being equal to 58 W/m^2 of body surface area (corresponding to the activity of sitting). The insulation value of clothing is measured in units known as the *clo*, one met being equal to 0.155 m²K/W. Values of these personal variables for some typical activities and types of clothing are given in Tables 3.2 and 3.3, respectively [5].

The comfort equation is complex, and its solution requires multiple iterations. Therefore, for direct application, diagrams have been prepared using computer analysis of the comfort equation for all the relevant combinations of the variables. Representative charts are reproduced in Fig. 3.2, the curved lines giving the optimum comfort condition, indicating the combination of conditions that satisfy the comfort equation [5].

Table 3.2 Heat production in the body at different typical activities clothing

Activity	Met
Sleeping	0.7
Sitting	0.8
Typing	1.1
Standing short sleeves and light underwear	1.4
Ordinary standing work in shop, laboratory, and kitchen	1.6–2.0
Slow walking (3 km/h)	2.0
Normal walking (5 km/h)	2.6
Fast walking (7 km/h)	3.8
Ordinary carpentry and brick-laying work	3.0

Table 3.3 Thermal resistance of different clothing ensembles

Clothing	Clo
Underpants plus	
○ Shirt (short sleeves), lightweight trousers, light socks, and shoes	0.5
○ Shirt, lightweight trousers, socks, and shoes	0.6
○ Boiler suit, socks, and shoes	0.7
Underwear (short sleeves/legs) plus	
○ Tracksuit (sweater and trousers), long socks, and training shoes	0.75
○ Boiler suit, insulated jacket and trousers, socks, and shoes	1.4
○ Shirt, trousers, jacket, quilted jacket and overalls, socks, and shoes	1.85
Bra and pants plus	
○ T-shirt, shorts, light socks, and sandals	0.3
○ Stockings, blouse (short sleeves), skirt, and sandals	0.55
○ Shirt, skirt, sweater, thick socks (long), and shoes	0.9

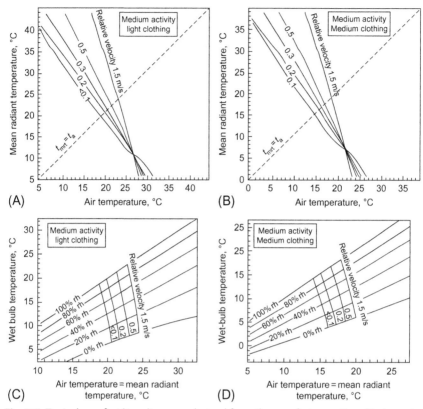

Fig. 3.2 Typical comfort line diagrams derived from the comfort equation. *(Redrawn by permission of McGraw Hill Book Co.)*

Example 3.1

In a small factory, the occupants are doing light bench work (medium activity). For both cases, determine the optimum air temperature, assuming that the mean radiant temperature is close to the air temperature, and that the air velocity is in the range of 0.1–0.2 m/s.

Solution

It is assumed that in summer light clothing (clo = 0.5) is worn and in winter, medium clothing (clo = 1.0).

From Fig. 3.2A, at an air velocity of 0.15 m/s:

Summer air temperature   20°C

From Fig. 3.2B, at an air velocity of 0.15 m/s:

Winter air temperature   15.5°C

At the optimum comfort condition, there is likely to be a level of dissatisfaction of about 10%, and as the conditions deviate from the optimum, the level of discomfort will increase, as indicated previously in Fig. 3.1. The comfort equation may also be used to examine the range of conditions that will be suitable for a particular design.

Example 3.2

For the conditions given in Example 3.1, determine the effect on the optimum dry-bulb temperature of relative humidity in the range of 30%–70%.

Solution
Summer

From Fig. 3.2C, at an air velocity of 0.15 m/s, dry-bulb temperatures are read as:

  at 30%rh = 20.5°C
  at 70%rh = 19.8°C

Winter

From Fig. 3.2D, at an air velocity of 0.15 m/s, dry-bulb temperatures are read as:

  at 30%rh = 16.0°C
  at 70%rh = 15.0°C

OTHER CONSIDERATIONS

The indoor air conditions that have to be maintained within an air conditioned space depend on a number of considerations other than overall thermal comfort.

Pleasant Indoor Environments

Creating a comfortable environment within an air conditioned space depends not only on obtaining the optimum *average* room conditions but also on attempting to produce within the room a *pleasant and refreshing* atmosphere. Conversely, local feelings of discomfort have to be avoided.

The term *freshness* is the most commonly used word to describe this aspect of comfort. The ventilation rate, usually the supply of outdoor air, required to reduce odours and other airborne contaminants to acceptable levels, plays an important role in creating a *fresh* atmosphere. Recommended ventilation rates for this aspect of design are given in Chapter 5. Freshness has also been associated with the ionization of the air. (See Croome–Gale and Roberts [6] for a discussion on this topic.)

Dry-bulb Temperature

In most buildings, the dry-bulb temperature is usually the most significant variable affecting comfort and therefore the most closely controlled. As important as the average, or controlled temperature, is the variation of air temperature in the space, especially the variations from floor to ceiling. An ideal environment is one in which the temperature at floor level is slightly greater than at head level, though in practice this is difficult to achieve with conventional air conditioning systems. Excessive vertical temperature gradients produce a feeling of *stuffiness* in the atmosphere; these temperature gradients are almost always associated with a lack of air movement in the occupied zone of the room, particularly at head height, this being brought about by inadequate air diffusion within the space. Almost certainly, the failure of many early air heating systems to produce satisfactory conditions was due to this effect. Bedford [7] suggested that the maximum temperature difference from floor to head height should not be greater than 3 K and should preferably be less than this.

Relative Humidity

Reference to Fanger's comfort diagrams and illustrated by Example 3.2 shows that relative humidity has little effect on the overall sensation of thermal comfort. A change of relative humidity of 40% can be compensated by approximately 1°C in dry-bulb temperature, and therefore, a wide tolerance of the room humidity can be accepted where comfort is the main consideration. However, a significant correlation has been found between humidity and freshness. Bedford [8] advised that, to create a pleasant and invigorating environment, the relative humidity should not exceed 70% and preferably be well below this figure. It was been suggested by Seeley [9] that drastic changes in moisture in the nasal cavity, which can occur with a sudden change in the air condition, should be avoided. This implies that the outdoor air conditions should be taken into account in the choice of indoor conditions, a warmer humid climate would require an indoor air condition that was relatively more humid than that of a warm, dry climate. Although a level of humidity may also be the optimum for a feeling of freshness, achieving that level in summer when cooling and dehumidifying are required will be more expensive than maintaining relative humidity at a higher level. Humidity is only one factor in the creation of a pleasant indoor atmosphere and if other criteria are met, then humidity becomes less important.

The buildup of static electricity on machinery and carpets often causes problems. For example, with printing presses an electrostatic charge on the press will cause the paper to stick to the machinery; in this case, the relative humidity should be at least 55%. For carpets, the relative humidity should be maintained above 40% unless underfloor heating is involved (causing very dry carpets) when it should be a minimum of 55%.

In air conditioned buildings, relative humidity below 40%rh increases the risk of sick building syndrome (SBS) [10].

To prevent condensation on internal surfaces and interstitial condensation within the building fabric, it may be necessary to limit the room humidity. A maximum of 60%rh is recommended to prevent bacterial growth.

Humidity levels may also have a significant effect on system energy consumption. It has been calculated that for some systems, the theoretical annual cost savings on refrigeration compressor energy consumption can be halved if the room relative humidity is allowed to rise to 60%rh compared with maintaining 50%rh [11].

Air Movement

It is generally well known that wind accompanied by low air temperatures produce a cooling effect on the body far greater than that caused by temperature alone. Daily weather forecasts in winter often quote *wind-chill* conditions, an index of cooling that combines temperature with wind speed.

Cold draughts in occupied spaces are also a well-known phenomenon, though more difficult to predict than the overall cooling effect of a low-temperature wind. A draught is a localized sensation, also caused by a combination of temperature and velocity of the airstream, felt mostly on the back of the neck and by the ankles. Warm draughts may also occur, but these do not present the designer of air conditioning systems with the same scale of problem as a cold draught. Unless the air diffusion is carefully planned, it is often difficult to avoid producing cold draughts at times of large cooling loads. To avoid the sensation, it is usual to design for air velocities in the occupied zone of the room in the range of 0.1–0.2 m/s.

An impression of freshness can be created by a variable air movement (small-scale air turbulence) promoted by means of the air diffusion system; the level of sophistication of the supply air outlets to achieve satisfactory results will be reflected in the capital cost of these items. With variable air volume systems, it is possible that less than satisfactory air diffusion will be obtained when the outlet is operating at low flow rates (see Chapter 5).

Mean Radiant Temperature

The temperatures of the surrounding surfaces have a significant effect on thermal comfort. Apart from the use of radiant heating/cooling systems, the MRT may not be directly controlled since it is a function of the internal fabric surfaces, radiation effects of the sun, and the warm surfaces of machines and lights. For new buildings with higher fabric insulation standards, the indoor surface temperatures in winter will be close to the air temperature, leading to the conclusion in this case that control of the air temperature will be satisfactory for comfort. However, in spring, summer, and autumn, the MRT will often be significantly higher than the air temperature, especially where heat-absorbing glass and/or internal blinds are used. Where the air dry-bulb temperature differs appreciably from the MRT, the air temperature may be adjusted through the control system to give an equivalent comfort sensation. For freshness, the surrounding wall surfaces should be uniformly at a somewhat higher temperature than the air temperature [12].

Space Conditions for Other Requirements

Other requirements for determining indoor space conditions include the operating efficiency of machinery, processes, and the preservation of artefacts; a compromise may therefore have to be struck between these conditions and those required for thermal comfort.

INDOOR DESIGN CONDITION ENVELOPES

In choosing indoor air design conditions, it is also necessary to decide on the allowable deviation from an optimum condition, in order that the amount of zoning and level of sophistication of the control system may be determined. The aim of the design engineer should be to satisfy the large majority of the occupants within the comfort zone; care should be taken not to specify precise values for the indoor conditions when a range of temperature and humidity would be equally acceptable, since this may incur unnecessary capital expenditure. Once these ranges have been agreed, they may be shown on a psychrometric chart.

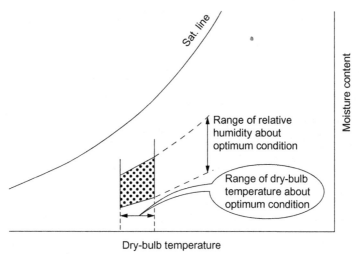

Fig. 3.3 Indoor design condition envelopes with a range of relative humidity.

A design condition *envelope*, for a room in which the control of air temperature and relative humidity are required, is shown in Fig. 3.3; the envelope in which the air temperature is maintained with the absolute humidity (or dew-point temperature) set at high and low limits, is shown in Fig. 3.3.

The AC plant should be designed and controlled to achieve the space conditions throughout the year, to a given tolerance, within the envelope (Fig. 3.4).

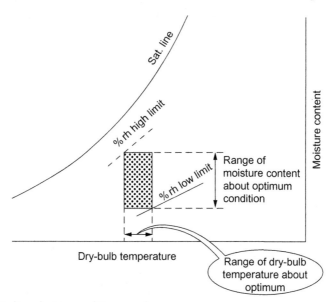

Fig. 3.4 Indoor design condition envelopes.

MEASUREMENT OF SPACE CONDITIONS

Comfort meters are available for evaluating the indoor environment. These instruments measure the four environmental variables, combining them according to the comfort equation to produce a single numerical value. Such instruments are possibly satisfactory for environments that are already comfortable(!), but when room conditions are less than satisfactory, it will be necessary to measure the individual variables. Suitable dry-bulb temperature and humidity instruments are described in Chapter 1.

Air velocities may be measured with a variety of instruments such as hot-wire anemometers; for the low air speeds associated with comfort air conditioning, a Kata thermometer is often used.

A globe thermometer is suitable for the measurement of the MRT; the instrument consists of a thermometer whose bulb is at the centre of a blackened, hollow sphere.

The globe temperature, t_g, measured with a 150 mm sphere, is related to the air temperature, t_a, velocity, v, and mean radiant temperature, t_r. According to the following equation,

$$t_g = \frac{t_r + 2.35\, t_a\, \sqrt{v}}{1 + 2.35\sqrt{v}}$$

SYMBOLS

t_a air temperature (dry-bulb)
t_g globe thermometer temperature
t_r mean radiant temperature
v air velocity

ABBREVIATIONS

clo   insulation value of clothing
met   heat produced by human body, metabolic rate
MRT   mean radiant temperature
SBS   sick building syndrome

REFERENCES

[1] CIBSE Guide A: Environmental Design, 2015, p 1.3.
[2] ASHRAE Standard 55: Thermal Environmental Conditions for Human Occupancy, 2013.
[3] T. Bedford, Industrial Health Research Board Report. Medical Research Council, (1936) 76, pp. iv + 102.

[4] D.E. Hickish, Thermal sensations of workers in light industry in summer: a field study in Southern England, J. Hyg. 53 (1955) 11.

[5] P.O. Fanger, Thermal Comfort, McGraw-Hill, New York, 1973.

[6] D.J. Croome-Gale, B.M. Roberts, Air Conditioning and Ventilation of Buildings, Pergamon Press, Oxford, 1981.

[7] T. Bedford, Requirements for satisfactory heating and ventilating, Ann. Occup. Hyg. 1 (1960) 172.

[8] T. Bedford, Basic Principles Heating and of Ventilation and Heating, Lewis, London, 1984.

[9] L.J. Seeley, Change in temperature of air in nasal cavity, Am. Soc. Heat. Vent. Eng. 46 (1940) 285.

[10] CIBSE Guide A, op. cit., p1.5.

[11] R.C. Legg, Analysis of energy demands of air conditioning systems, Heat. Vent. Eng. 51 (1977) 6.

[12] T. Bedford, Requirements for satisfactory heating and ventilating, Ann. Occup. Hyg. 2 (1960) 167.

CHAPTER 4

Outdoor Air Design Conditions

Suitable outdoor air conditions should be selected for the design of air conditioning systems so that the maximum plant loads can be determined; this applies to both summer and winter system load requirements. The designer, making an inappropriate choice, may either select conditions that are too extreme, leading to either:

- oversized plant;
- inflated capital cost;
- poor plant efficiency;
- increased operating costs;
- poor plant control.

or select conditions that are insufficiently stringent, leading to:

- undersized plant;
- indoor design conditions not being maintained at times of peak load.

FREQUENCY OF OCCURRENCE OF OUTDOOR AIR CONDITIONS

The Meteorological Office of the United Kingdom has published a set of reports dealing with the combined frequency distribution of dry-bulb and screen wet-bulb temperature (WBT) for 28 stations throughout the United Kingdom [1]. A number of analyses have been made of these data so that they should be presented in a form suitable for use by air conditioning engineers [2]. Thus, percentage frequency distributions were obtained for the following air properties:

  (i)   dry-bulb temperature (DBT);
  (ii)   wet-bulb temperature (WBT);
 (iii)   specific enthalpy;

(iv) DBT in association with moisture content.

These frequency distributions, for the whole of the annual 8760 hourly observations, were for the following daily periods:

(a) 24 h
(b) 12 h   From 07.00 to 18.00 GMT
     12 h   From 19.00 to 06.00 GMT
(c) 8 h   From 02.00 to 09.00 GMT
     8 h   From 10.00 to 17.00 GMT
     8 h   From 08.00 to 01.00 GMT

Typical 24 h frequency distributions for Heathrow for DBT, WBT, and specific enthalpy are given in Tables 4.1–4.3, respectively. In each of these tables, the frequency of occurrence of the measured air condition, for a range of the air property, is given as a percentage of the total hours in the year; for DBT in association with moisture content, the data are plotted on a psychrometric chart in Fig. 4.1. In this figure, each 'box' contains the (percentage $\times$ 100) of the measured air conditions that were within the limits of 2 K DBT and 0.001 kg/kg_{da} moisture content.

The values given as 0.00% in Tables 4.2–4.4 (zero in Fig. 4.1) are the extreme conditions recorded with a frequency of less than 0.005%.

Summer Design Conditions from Frequency Tables

To determine a design condition for summer, a target of a total accumulated percentage (percentile) of annual hours is set, at which the outdoor air condition is reached or exceeded; this percentile is a considered judgment made by the design engineer. Individual frequencies of the appropriate air property are then summed from the extreme condition until the target is reached, within reasonable accuracy.

Table 4.1 Percentage frequency distribution of hourly values of outdoor air dry-bulb temperature, annual 24 h periods, Heathrow, the United Kingdom (percentage × 100)

Dry-bulb temperature 1 K intervals (°C)	Frequency f_1 (%)	Dry-bulb temperature 1 K intervals (°C)	Frequency f_1 (%)	Dry-bulb temperature 1 K intervals (°C)	Frequency f_1 (%)
		0.0–0.9	2.19	20.0–20.9	1.71
		1.0–1.9	2.78	21.0–21.9	1.27
		2.0–2.9	3.30	22.0–22.9	0.98
		3.0–3.9	3.85	23.0–23.9	0.63
		4.0–4.9	4.16	24.0–24.9	0.46
		5.0–5.9	5.12	25.0–25.9	0.31
< −13.1	—	6.0–6.9	5.46	26.0–26.9	0.20
−13.0 to −12.1	0.00	7.0–7.9	5.82	27.0–27.9	0.12
−12.0 to −11.1	0.00	8.0–8.9	6.21	28.0–28.9	0.07
−11.0 to −10.1	0.01	9.0–9.9	6.01	29.0–29.9	0.06
−10.0 to −9.1	0.01	10.0–10.9	6.08	30.0–30.9	0.04
−9.0 to −8.1	0.01	11.0–11.9	5.76	31.0–31.9	0.02
−8.0 to −7.1	0.02	12.0–12.9	5.63	32.0–32.9	0.02
−7.0 to −6.1	0.04	13.0–13.9	5.47	33.0–33.9	0.01
−6.0 to −5.1	0.08	14.0–14.9	5.08	34.0–34.9	0.00
−5.0 to −4.1	0.16	15.0–15.9	4.97	>35.0	—
−4.0 to −3.1	0.30	16.0–16.9	4.40		
−3.0 to −2.1	0.52	17.0–17.9	3.57		
−2.0 to −1.1	0.83	18.0–18.9	2.82		
−1.0 to −0.1	1.38	19.0–19.9	2.09		

Table 4.2 Percentage frequency distribution of hourly values of outdoor air wet-bulb temperatures (sling), annual 24 h periods, Heathrow, the United Kingdom (percentage × 100)

Wet-bulb temperature 1 K intervals (°C)	Frequency f_1 (%)	Wet-bulb temperature 1 K intervals (°C)	Frequency f_1 (%)	Wet-bulb temperature 1 K intervals (°C)	Frequency f_1 (%)
< −13.1	—	0.0–0.9	2.77	20.0–20.9	0.13
−13.0 to −12.1	0.00	1.0–1.9	3.46	21.0–22.9	0.04
−12.0 to −11.1	0.00	2.0–2.9	4.18	22.0–22.9	0.01
−11.0 to −10.1	0.01	3.0–3.9	4.76	>923.0	—
−10.0 to −9.1	0.01	4.0–4.9	5.63		
−9.0 to −8.1	0.01	5.0–5.9	5.89		
−8.0 to −7.1	0.03	6.0–6.9	6.47		
−7.0 to −6.1	0.06	7.0–7.9	6.85		
−6.0 to −5.1	0.11	8.0–8.9	6.78		
−5.0 to −4.1	0.23	9.0–9.9	6.94		
−4.0 to −3.1	0.47	10.0–10.9	6.98		
−3.0 to −2.1	0.80	11.0–11.9	6.67		
−2.0 to −1.1	1.38	12.0–12.9	6.72		
−1.0 to −0.1	2.29	13.0–13.9	6.39		
		14.0–14.9	5.36		
		15.0–15.9	3.98		
		16.0–16.9	2.77		
		17.0–17.9	1.38		
		18.0–18.9	0.70		
		19.0–19.9	0.35		

Table 4.3 Percentage frequency distribution of hourly values of outdoor air specific enthalpy, annual 24 h periods, Heathrow, the United Kingdom (percentage × 100)

Specific enthalpy 2 kJ/kg$_{da}$ intervals	Frequency f_3 (%)	Specific enthalpy 2 kJ/k g$_{da}$ intervals	Frequency f_3 (%)
−10.0 to −8.1	0.00	30.0–31.9	5.58
−8.0 to −6.1	0.01	32.0–23.9	5.54
−6.0 to −4.1	0.01	34.0–35.9	5.36
−4.0 to −2.1	0.03	36.0–37.9	5.06
−2.0 to −0.1	0.07	38.0–39.9	4.40
0.0–1.9	0.20	40.0–41.9	3.72
2.0–3.9	0.46	42.0–43.9	2.99
4.0–5.9	0.92	44.0–45.9	2.16
6.0–7.9	1.67	46.0–47.9	1.42
8.0–9.9	2.98	48.0–49.9	0.85
10.0–11.9	3.45	50.0–51.9	0.51
12.0–13.9	4.33	52.0–53.9	0.33
14.0–15.9	4.98	54.0–55.9	0.18
16.0–17.9	5.43	56.0–57.9	0.10
18.0–19.9	6.31	58.0–59.9	0.05
20.0–21.9	6.18	30.0–31.9	0.02
22.0–23.9	6.45	32.0–23.9	0.01
24.0–25.9	6.23	34.0–35.9	0.00
26.0–27.9	6.18	>66.0	—
28.0–29.9	6.01		

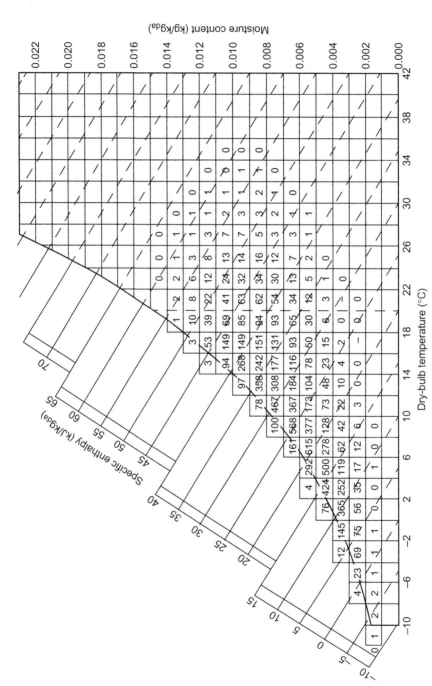

Fig. 4.1 Percentage frequency distribution of hourly values of outdoor air dry-bulb temperature in association with moisture content, annual 24 h periods, Heathrow, the United Kingdom (100 ×).

Example 4.1

Determine the outdoor design conditions for an air conditioning system for a building near Heathrow, London. A 99.65% percentile (equivalent to the design condition being exceeded for 29 h per year) is considered appropriate.

Solution

To obtain the summer design conditions, the frequencies are summed commencing from the maximum property in the table until the target of 99.65%–100% is reached. All 0.00% values are ignored. The calculations are set out in the tables below.

Dry-bulb Temperature

Refer to Table 4.1 and commence at the extreme value of 34°C:

Dry-bulb temperature 1 K intervals (°C)	Frequency f_1 (%)
27.0–27.9	0.12
28.0–28:9	0.07
29.0–29.9	0.08
30.0–30:9	0.04
31.0–31.9	0.02
32.0–32.9	0.02
33.0–33.9	0.01
$\Sigma(f_1)$	0.34

Therefore, the design DBT is 27°C.

Wet-bulb Temperature (Sling)

Refer to Table 4.2 and commence at the extreme value of 23°C:

Wet-bulb temperature intervals (°C)	Frequency f_2 (%)
19.5–19.9	0.17
20.0–20:9	0.12
21.0–21.9	0.04
22.0–22.9	0.01
$\Sigma(f_1)$	0.35

Therefore, the design WBT is 19.5°C.

Specific Enthalpy

In the same way, a design specific enthalpy can be obtained using Table 4.3, commencing at the extreme value of 64 kJ/kg$_{da}$; with $\Sigma(f_3) = 0.36$, a design condition of 54 kJ/kg$_{da}$ is obtained. The summer design conditions for the air conditioning system(s) will therefore be as follows:

Dry-bulb temperature	27°C
Wet-bulb temperature (sling)	19.5°C
Specific enthalpy	54 kJ/kg$_{da}$

Note that, due to rounding errors, the three conditions will not coincide precisely.

Winter Design Conditions

Similar calculations can be made for winter design conditions. In this case, a target percentile is set, in which the air property is at or below the design condition. The individual frequencies of the air property being considered are then summed from the extreme minimum condition until the percentile is reached.

Example 4.2

Determine the outdoor design conditions for an air conditioning system that is being specified for a building near Heathrow, London. A percentile of 0.35% (equivalent to 29 h per year) is considered suitable.

Solution

To obtain the winter design conditions, the frequencies are summed commencing from the minimum property in the table until the target percentile of 0.35% is reached. All 0.00% values are ignored. The calculations are set out in the tables below.

Dry-bulb Temperature

Refer to Table 4.1 and commence at the extreme value of −11°C in the following:

Dry-bulb temperature 1 K intervals (°C)	Frequency f_1 (%)
−11.0 to −10.1	0.01
−10.0 to −9.1	0.01
−9.0 to −8.1	0.01
−8.0 to −7.1	0.02
−7.0 to −6.1	0.04
−6.0 to −5.1	0.08
−5.0 to −4.1	0.16
$\Sigma(f_1)$	0.33

Therefore, the design DBT is −4.5°C.

Wet-bulb Temperature

Refer to Table 4.2 and commence at the extreme value of $-11°C$ in the following:

Dry-bulb temperature 1 K intervals (°C)	Frequency f_1 (%)
−11.0 to −10.1	0.01
−10.0 to −9.1	0.01
−9.0 to −8.1	0.01
−8.0 to −7.1	0.03
−7.0 to −6.1	0.06
−6.0 to −5.1	0.11
−5.0 to −4.5	0.11
$\Sigma(f_1)$	0.34

Therefore, the design WBT is $-4.5°C$.

Specific Enthalpy

In the same way, a design specific enthalpy can be obtained using Table 4.3, commencing at the extreme value of -8 kJ/kg_{da}; with $\Sigma(f_3) = 0.32$, a design condition of 2 kJ/kg_{da} is obtained. The winter design conditions for the air conditioning system(s) will be

Dry-bulb temperature	$-4.0°C$
Wet-bulb temperature	$-4.5°C$
Specific enthalpy	2 kJ/kg_{da}

The choice of which air properties to use as the design condition(s) will depend on the application. For example, DBT will be used for sensible heating processes, WBTs for cooling tower selection, and enthalpy for dehumidifying cooling coils.

Design Conditions for the United Kingdom

The CIBSE Guide B [3] tabulates warm temperatures for 14 cities in the United Kingdom; these are reproduced in Table 4.4. The DBTs are those equalled or exceeded for the given percentages of hours for the years 1982–2011 (Leeds, 1989–2011, and Newcastle, 1983–2011). The WBTs are the average coincident temperature associated with the DBTs. The percentages are equivalent to 35, 88, 175, and 438 h a year. It should be noted that these data are based on historical records, and with global warming, they are likely to exceed the temperatures given in the table. These temperatures can be used for selecting summer outdoor design conditions.

Table 4.4 Warm temperatures in the United Kingdom

| Location | Temperatures (°C) equal to or exceeded for given percentage | | | | | | | |
| | 0.4% | | 1% | | 2% | | 5% | |
	DBT	WBT	DBT	WBT	DBT	WBT	DBT	WBT
Belfast	22.5	17.5	20.7	16.5	19.2	15.7	17.3	14.5
Birmingham	26.3	18.9	24.2	17.7	22.5	16.9	19.8	15.5
Cardiff	24.4	18.2	22.5	17.3	20.8	16.5	16.6	15.5
Edinburgh	23.3	17.1	20.7	16.2	19.4	15.4	17.4	14.1
Glasgow	25.9	17.6	21.3	16.5	19.6	15.6	17.4	14.3
Leeds	25.9	18.4	24.0	17.5	22.2	16.6	19.7	15.3
London	28.1	19.2	261	18.3	24.4	17.6	21.6	16.2
Manchester	25.5	18.4	23.4	17.4	21.7	16.5	19.1	15.2
Newcastle	23.0	17.2	21.4	16.3	19.9	15.5	17.8	14.2
Norwich	26.6	19.2	24.7	18.3	22.9	17.4	20.3	15.9
Nottingham	26.0	19.2	24.0	17.7	22.1	16.8	19.4	15.3
Plymouth	23.4	17.9	21.7	17.1	20.3	16.5	18.4	15.6
Southampton	25.7	19.0	23.9	17.9	22.4	17.1	20.0	16.0
Swindon	26.8	18.8	24.8	17.8	23.0	16.9	20.2	15.5

Data reproduced from CIBSE Guide A with the permission from the Chartered Institute of Building Services Engineers.

These data may be used to determine outdoor air design conditions, system operating requirements at off-peak conditions and annual energy consumption (see Chapter 18). Calculations using these tables do not have to be precise; the nearest 0.5 K is sufficient for temperatures and 1.0 kJ/kg$_{da}$ for enthalpies. Though the frequencies of DBT, WBT, and specific enthalpy are not coincident with each other, they can be considered to be self-consistent, since the air properties were derived from the same set of measurements.

WORLD-WIDE DATA

The ASHRAE Handbook [4] tabulates outdoor air conditions for over 6000 locations world-wide. The recommended summer design conditions are based on 1%, 2.5%, and 5% total hours in which the DBTs are reached or exceeded in the four summer months Jun.–Sep. (Dec.–Mar. in southern hemisphere). These percentages are equivalent to 29, 73, and 146 h per annum, respectively. The coincident WBTs listed with each DBT are the *mean* of all the WBTs occurring at the specific DBT and therefore not as stringent as the WBTs computed independently of the DBT. For the United Kingdom, there are 26 sites (compared with the CIBSE Guide's 14 sites).

Recommended winter conditions of DBT are based on 99% and 97.5% criteria. Both frequency levels represent the total hours in the three winter months Dec., Jan., and Feb. (Jun., Jul., and Aug. for southern hemisphere), during which the temperature levels were reached or exceeded. That is to say, those temperatures are at or below the conditions given, for an average of 22 and 54 h, respectively, in those three months.

OUTDOOR AIR CONDITION ENVELOPES

Air conditioning systems will usually be required to operate throughout the year against a whole range of outdoor air conditions. The extent of these conditions is most usefully shown on an *outdoor air condition envelope*, based on the frequency distributions similar to that shown in Fig. 4.1.

These envelopes provide a useful tool for summarizing plant operations; e.g., see Chapter 6. The design conditions of DBT and specific enthalpy, determined in the previous example, are shown on the skeleton envelopes in Figs 4.2 and 4.3, respectively; the shaded areas on each envelope represent the *total* accumulated frequency in which the outdoor air condition falls above, or below, the design condition.

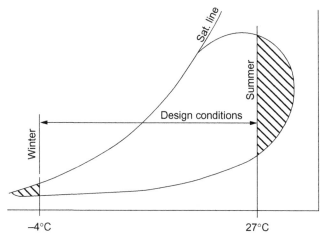

Fig. 4.2 Outdoor air condition envelope with design dry-bulb temperatures.

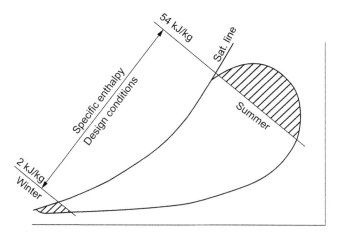

Fig. 4.3 Outdoor air condition envelope with design specific enthalpies.

Points to Note

- The design conditions are an average based on historical data; it is most unlikely that, for any 1 year, these conditions would actually occur at the frequency predicted.
- Most of the published meteorological records were obtained on airfields or similarly exposed sites. Air conditions at a proposed site may differ significantly from those records, and therefore, the local meso- and microclimate must be considered if realistic conditions are to be obtained

(e.g., cities act as *heat islands*). A similar comment applies to the interpolation of data for a location that is between data stations.

- Care should be taken in positioning the fresh-air intake for the air conditioning system. Ambient conditions can be affected by solar produced convection currents on some faces of the building, heat gains from the plant room and from air discharges from exhaust systems.
- The outdoor design conditions will not be critical if heat recovery from the exhaust air is employed in systems using 100% outdoor air or with systems employing a large percentage of recirculated air.

REFERENCES

[1] Meteorological Office (UK), Climatological Memorandum, frequencies of hourly values of dry-bulb temperature and associated wet-bulb temperature. Various dates.
[2] R.H. Collingbourne, R.C. Legg, The frequency of occurrence of hourly values of outside air conditions in the United Kingdom, IoEE Technical Memorandum No 58, South Bank Polytechnic, 1979.
[3] CIBSE Guide A: Environmental Design 2015.
[4] ASHRAE Handbook Fundamentals, Chapter 14, 2013.

CHAPTER 5

Room Heat Gains, Air Diffusion, and Air Flow Rates

With the indoor and outdoor design conditions selected, the design process then requires the calculation of the summer heat gains and winter heat losses. To maintain satisfactory indoor conditions, air supply and extract flow rates are determined based on a number of considerations, including the need to offset these gains and losses. This in turn requires an understanding of the way the air is supplied and extracted from the space through outlet grilles and diffusers so that a pleasant, draught-free environment is achieved. Air flow rates must also meet ventilation and air movement control requirements of the occupied building.

HEAT GAINS

The ideal structure, from the thermal point of view, would be one that modified the gains so that the temperatures are within the comfort zone. Since buildings are rarely built like this, either the excess gain has to be dealt with by mechanical cooling and/or ventilation or the temperature is allowed to rise, causing discomfort.

Sensible heat gains to the air conditioned space arise from the following sources:
- Solar radiation through windows;
- Solar radiation on the outside surface of the building structure (walls and roofs);
- Heat transmission through the building fabric due to outdoor/indoor temperature differences;
- Infiltration of outdoor air;
- Occupants;
- Lighting;
- Machines;
- Processes.

The total heat gains from these sources is the *design sensible heat gain*, q_s, which becomes the cooling load for the space. To obtain an accurate estimate of total gain, care must be taken when adding the individual sources together.

Air Conditioning System Design
http://dx.doi.org/10.1016/B978-0-08-101123-2.00005-4

Diversity factors and time lags should be applied where appropriate; if two or more of the gains vary through the day and are out of phase, then they should be added according to the time at which they occur.

Storage Effect of the Building Structure

The effect of heat gains that are cyclic in nature, such as the gains through windows, is to reduce the *instantaneous* heat gain through the storage effect of the structure. This is illustrated in Fig. 5.1. The net heat gain to the space equals the cooling load, which has to be dealt with by the air conditioning system.

Two classifications of structure are used for air conditioned buildings. These are the following:

- *Heavyweight building*—a building with solid floors, ceilings and solid internal walls, and ceilings;
- *Lightweight building*—a building with lightweight demountable partitions and suspended ceilings. Floors are either solid with carpet or wood block finish or suspended type.

The cooling load arising from an instantaneous sensible heat gain will be reduced more by the former than by the latter. If the indoor temperature is allowed to rise (swing) above the design value during peak periods, the plant size can be reduced by assessing the reduced cooling load for the space.

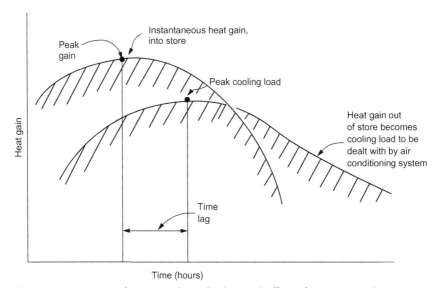

Fig. 5.1 Instantaneous beat gain through glass and effect of structure in determining the air cooling load.

REDUCTION OF HEAT GAINS

Some sources of heat gain can be excessive, imposing large loads on the air conditioning systems. While details of the calculation methods are not given here, it is often important that architects and engineers should consider ways of reducing heat gains to the building; by limiting the heat gains at source, plant sizes can be reduced. The methods of reducing gains include the following:

Windows

Solar heat gains through windows are often the main source of excessive cooling loads. However, gains through windows can be useful in offsetting heat loss in winter and thus reducing energy demands. A balance is required to optimize the window size/shading device to limit the maximum gain in summer and the maximum heat loss in winter, so that the overall plant sizes and operating costs can be reduced. Another important factor in the optimization of window size is the lighting requirement. Good daylighting can reduce the use of artificial lighting and hence limit heat gains and energy consumption.

Solar heat gain through glazing may be reduced by one or more of the following:
- Limiting window area;
- External solar shading devices;
- Internal solar shading devices;
- Blinds between panes of double glazing;
- Various types of glass;
- Orientation.

Glazing Area

Probably, the most satisfactory way to reduce the solar heat gain through glazing is to limit the window area in relation to the outside wall. Hardy and Mitchell [1] give some guidance to reasonable glazing areas as shown in Table 5.1.

Table 5.1 Recommended maximum glazing area as a percentage of external wall area

Building construction	Maximum % of glazing	
	Air conditioning	Mechanical ventilation
Heavyweight	70	45
Lightweight	50	20

Shading Devices

Shading devices on the external side of the window include shutters, awnings, canopies, blinds, and projecting horizontal and vertical fins. Correctly designed, these sun controls are the most effective of all for reducing solar radiation since the absorbed heat is dissipated externally. These devices, which can be fixed, adjustable or retractable, have to be designed to prevent direct radiation falling on the window at appropriate times of the day and year. Fixed projections are most suitable in the tropics and subtropics where the sun's altitude is high but less effective in temperate regions such as Britain. This is because they would have to project a distance further than the window height to give adequate protection at low sun angles. In so doing, daylight would be reduced and possibly impair the visual environment. Balconies have been used to good effect in some buildings, a notable example being Guy's Hospital tower in London, a detail of which is shown in Fig. 5.2. (The balcony also assists with window cleaning.) Adjustable external louvers can have high maintenance costs; manual operation is impractical if the building is more than two stories high.

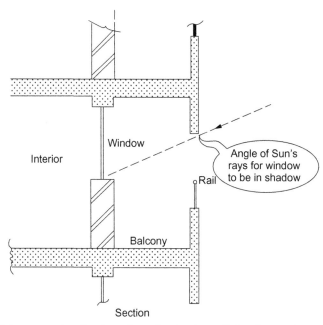

Fig. 5.2 Balcony construction to reduce direct solar radiation heat gains.

Internal Blinds

Internal blinds can give adequate protection against glare and direct heating effects of sunshine, but they are not very effective in reducing solar heat gain. This is because most of the radiant heat is absorbed by the blind together with an increase in the MRT. As blinds are of small mass, these effects take place fairly quickly. To give most benefit, the blinds should have a white polished surface facing the window to reflect as much radiation as possible. Dark colour blinds and curtains are least effective because they are better absorbers of radiant heat.

Blinds can be placed between the panes of double-glazed windows. In this position, the blinds are protected from dirt and mechanical damage. Where a double window, with panes 160–200 mm apart, is installed for the reduction of external noise, this solution might be the most satisfactory, but the MRT can rise because heat is trapped between the two panes of glass.

Types of Glass

More radiation is absorbed by thick glass, double, or triple glass. Special heat-absorbing glasses are available, which absorb infrared radiation without greatly reducing the transmission of light. The types of glass available transmit between 20% and 60% of solar radiation, but about 30% of the absorbed heat is retransmitted as convective heat and raising the MRT. The factors for the reduction of heat gain through a single pane of these glasses vary from 0.4 to 0.8. These can be reduced between 0.3 and 0.6 if the heat-absorbing glass is used as the external pane of a double-glazed window; then, the MRT will be reduced considerably.

Walls and Roofs

Heat gains through the opaque fabric of walls and roofs are likely to be small compared with gains from other sources, the main exception being the heat gains through a flat roof. To reduce heat gains through flat roofs, attention should be paid to the mass of the structure and to the thermal insulation. The outer surface should be finished with light-colour-reflecting material such as white gravel chips. Roofs are also sometimes sprayed with water in countries with high sun altitudes.

Infiltration of Outdoor Air

Any window that can be opened should be well fitting to minimize the infiltration of outdoor air. Special attention to the fabric of high-rise buildings may be required if excessive infiltration is to be prevented.

Lighting

A number of schemes have used extract air lighting fittings to reduce the heat gain to the space. These fittings are mounted in a false ceiling, which may then be used to provide an extract void. Though up to 80% of lamp heat can be removed, the *effective* heat removed is likely to be only approximately 50% since the warm extract air heats the ceiling from which heat is radiated to the space. A system using extract air lighting fittings is often designed so that extracted heat is made available as a source of heat in an application such as air-to-air heat recovery.

AIR DIFFUSION

Air diffusion is defined as the distribution of air in a treated space, using air terminal devices to satisfy the specified design conditions such as air change rate, pressure, cleanliness, temperature, humidity, air velocity, and noise level, in a space within the room termed the *occupied zone*. The occupied zone is usually taken to be 1.8 m in height above the floor and 0.3 m from the walls. (These dimensions do not apply in the case of either desk top or local air diffusion.) Air velocities within the occupied zone should be between 0.10 and 0.20 m/s.

Cooling Mode

To deal with heat gains, air is introduced into a space in a manner that will not cause draughts or excessive noise levels. For most designs, cooling air is supplied at temperatures well below and, at velocities well above, those that cause a sensation of draught; therefore, air must be supplied into a region of the room outside the occupied zone; this region is known as the *mixing zone*. The cold supply air introduced into this zone mixes with room air to form a *total airstream*. The velocities and temperature differences within the envelope of this total airstream reduce so that air can move into the occupied zone with little risk of causing draughts.

These general principles of air diffusion are illustrated in Fig. 5.3A for three traditional positions of the supply outlet, i.e., high-level wall grilles, ceiling diffusers, and window sill units. Details will vary with different grille types, location, room geometry, and use.

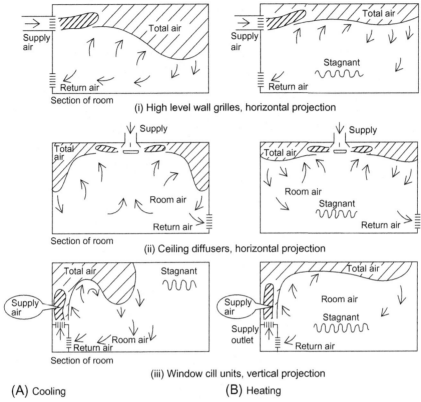

(A) Cooling   **(B)** Heating

Fig. 5.3 Room air diffusion patterns for typical supply outlet positions.

Heating Mode

To deal with heat losses in winter, air is supplied at a higher temperature than the controlled room condition. The warm supply air, introduced into the mixing zone, mixes with room air to form a total air envelope within which the velocity and supply-to-room temperature difference reduce. There is little danger of producing a warm draught, but if the temperature difference between the supply and room air is too great, the total air will remain close to the ceiling due to buoyancy effects. This may lead to temperature gradients from floor to ceiling and stagnant air in the occupied zone. The air distribution patterns for high-level wall grilles, ceiling diffusers, and window sill are illustrated in Fig. 5.3B.

COMMON TERMS USED IN AIR DIFFUSION

Supply air. It is the air flow rate supplied to a terminal device by the upstream duct. With some terminal devices, the total air delivered to the treated space by the device may be greater than the supply air because of air induced from the room into the device by virtue of its design.

Total air. It is the mixture of supply and induced room air, which is still under the influence of the supply air.

Throw. For a supply air outlet, it is the distance between the outlet and a plane, which is tangential to a specified envelope of air and perpendicular to the intended direction of flow (see Fig. 5.4). Outlet performance is often given in terms of the throw, usually with a terminal velocity of about

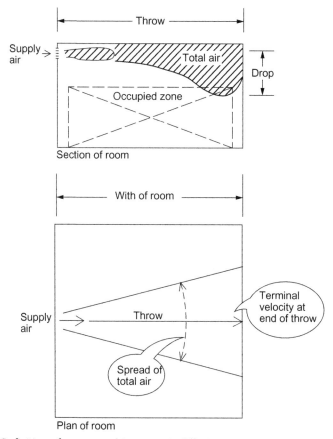

Fig. 5.4 Definition of terms used in room air diffusion.

0.5 m/s; acceptable air movement is achieved from an outlet expressed in terms of throw.

Terminal velocity. It is the velocity of the air stream at the end of its throw. The throw is usually taken as the horizontal distance across the room from the supply outlet. The terminal velocity is an arbitrary value that may range from 0.4 to 1.0 m/s for commercial applications and up to 2 m/s for industrial applications. For a given outlet, the throw is usually specified with the terminal velocity (see Fig. 5.4).

Overblow. It occurs when a jet of air from a terminal device meets an opposing wall and is deflected downward towards the occupied zone.

Drop. For a cold air supply dealing with a heat gain, if the room-to-supply temperature differential is too great or the throw is inadequate, air may 'drop' into the occupied zone, causing draughts (see Fig. 5.4).

Supply-to-room temperature differential (Δt). It is the difference between the supply air temperature and the mean temperature of the occupied space.

Stagnant region. It is a local region within the air conditioned space in which there is no measurable air velocity (or velocities less than 0.05 m/s).

Convection air currents. Variations in surface temperatures in a room give rise to localized heating or cooling of the room air. This sets up localized convection air currents in the space that may be caused by:

- Hot/cold windows;
- Hot/cold walls;
- Occupants;
- Lights and equipment;
- Radiators;
- Chilled panels.

Although difficult to quantify, natural circulation currents can significantly modify the pattern of air introduced through a grille.

Grille noise. The *CIBSE Guide* gives recommended maximum grille face velocities for tolerable room noise levels (see Table 5.2) [2] though these values will be modified by the acoustic properties of the room. However,

Table 5.2 Air velocities for acceptable noise levels

| Type of opening | Permitted air velocity for stated type of space (m/s) | | |
	Critical	Normal	Noncritical
Supply	1.5	2.5	3.0
Extract	2.0	3.0	4.0

it will be more usual to use the acoustic data prepared by the grille manufacturer where this is available.

OUTLET GRILLES AND DIFFUSERS

There are five basic types of supply outlets:
- Wall grille;
- Slot diffuser;
- Ceiling diffuser;
- Perforated ceiling panel;
- Displacement types.

Each of these has different performance characteristics. An important factor in the choice of outlet is its ability to entrain and mix room air, a relationship known as the induction ratio. The higher the induction ratio, the more air can be introduced into an air conditioned space without causing draughts, and the more quickly room air is mixed with the supply air. This implies that grilles with high induction ratio can handle larger temperature differences between the supply and room air. Ceiling diffusers have high induction ratios because they diffuse air in all directions. This leads to a short throw and more rapid temperature equalization than slot diffusers. Wall grilles have relatively long throws but low induction ratios.

GENERAL RULES FOR POSITIONING GRILLES AND DIFFUSERS

Supply Outlets
- If the outlet is to handle both warm and cool air at different times of the year, a compromise must be made in the choice of size and position to give satisfactory conditions throughout the year. As cold draughts are considered to be more uncomfortable than hot draughts, the designer should incline towards giving more satisfactory room distribution in summer when handling chilled air.
- Avoid overlap between the zones of influence of adjacent supply outlets, as an area of localized increased velocity can occur where two airstreams meet, giving rise to draughts.
- When supplying chilled air, an excessive drop is likely to occur if the outlet velocity is below 2 m/s.

- Variable-air-volume systems are likely to have a variable outlet velocity that will give rise to varying flow patterns in a room. The minimum acceptable outlet velocity is about 2 m/s; variable area outlet diffusers are available to reduce the outlet area and so maintain constant outlet velocities for satisfactory air diffusion.

  For upward vertical projection of supplying air from a floor or window sill-mounted outlet, the centre-line velocity should be reduced to approximately 0.75 m/s at ceiling level to avoid excessive downward deflection from the ceiling.
- The outlet(s) should be positioned to avoid supply air discharging against ceiling obstructions such as surface-mounted light fittings and downstand beams. Otherwise, some of the cold air will be deflected downward into the occupied space, causing draughts.

Exhaust Outlets

Exhaust air is the air that is either extracted or discharged from the conditioned space. This may take one or more of the following forms:
- *Extract*—exhaust in which the air is discharged to atmosphere;
- *Relief*—exhaust where air is allowed to escape from the treated space should the pressure in the space rise above a specified level;
- *Recirculation*—exhaust where air is returned to the air handling system;
- *Transfer*—exhaust lets air pass from the treated space to another treated space.

The zone of localized high velocity in association with extract grilles is very close to the grille, and therefore, the position of the extract grille has little influence upon the overall flow patterns in a room. For this reason, there is no need to match every supply outlet with an extract grille. The number of extracts is determined by physical limitations, together with the planning of the control of air movement through various rooms of the building, e.g., smoke control in the event of fire, clean and dirty areas of a hospital operating theatre suite of rooms. Extract grilles could be sited successfully at the following places:
- In a stagnant zone;
- Close to an excessive heat source, to minimize heat gain to the room;
- Close to an excessive cold source to minimize heat loss in the room;
- At a point of local low pressure, such as the centre of a circular ceiling diffuser discharging horizontally;

Exhaust grilles should *not* be positioned:

- in the zone of influence of a supply grille so as to prevent conditioned air passing directly into the extract system without having first exchanged heat with the surroundings;
- close to a door or aperture, which is frequently opened, so that the exhaust grille does not handle air from another space.

AIR VOLUME FLOW RATE

The supply and extract air volume flow rates required for each room or space in an air conditioned building are determined by a consideration of one or more of the following:

- Sensible heat gains;
- Latent heat gains;
- Sensible heat losses;
- Air distribution;
- Ventilation;
- Air movement control;
- Fire precautions.

Surveys of existing air conditioned buildings have shown that fan energy costs account for up to half the total energy costs of the system(s). Therefore, where it is important to reduce energy costs, it is important to design the systems at the lowest flow rates consistent with good design. Lower flow rates will also help to reduce the size of other energy-using equipment such as coolers and heaters.

Air Supply for Summer Cooling

Consider the air cooling to offset the design sensible heat gain q_s. The balance of the heat gain to the cooling effect of the air supply is given by Eq. (2.20), i.e.,

$$q_s = \dot{m}_a\, c_{pas} \left(t_R - t_S \right)$$

The design mass air flow rate is therefore given by:

$$\dot{m}_a = \frac{q_s}{c_{pas}\, \Delta t_c} \tag{5.1}$$

where $\Delta t_c = (t_R - t_S)$, the design cooling air temperature differential.

The supply air volume is obtained by using the specific volume of the supply air condition, v_S:

$$\dot{V} = \dot{m}_a v_S \qquad (5.2)$$

Since the room temperature t_R is the controlled condition, the flow rate is determined by setting the supply air temperature t_S. For economic design, the flow rate should be as small as possible, which means having a supply temperature as low as possible without causing draughts. Suitable temperature differentials depend on several factors, some of which have been discussed earlier in the chapter.

Example 5.1

A room is maintained at 21°C and has a design sensible heat gain of 6 kW. Determine the supply air volume flow rate if the room-to-supply air temperature differential is 8 K.

Solution

The supply air mass is obtained using Eq. (5.1):

$$\dot{m}_a = \frac{q_s}{c_{pas}\,\Delta t_c} = \frac{6}{1.02 \times 8} = 0.735\,\text{kg/s}$$

Using an average value of specific volume at the supply temperature of 13°C, the supply air volume is obtained from Eq. (5.2):

$$\dot{V} = \dot{m}_a v_S = 0.735 \times 0.82 = 0.602\,\text{m}^3/\text{s}$$

A number of other considerations are involved in the selection of a suitable temperature differential, some of which are examined with reference to the psychrometric sketch in Fig. 5.5.

Initially, a room condition **R** may be taken at the highest enthalpy in the indoor design envelope. From this, the room ratio line, **RX**, (with a relatively small slope for this particular example) is plotted. Any supply condition on this line will maintain the room condition against the design heat gains. The air leaving the cooling coil is at condition **B**, and **BS** is a temperature rise due to fan and duct heat gains. Thus, the maximum temperature differential Δt_c is determined by the position of the room ratio line on the chart, since it is limited by the saturation line and plant operating characteristics such as the contact factor of the coil. By lowering the room condition to R', a lower supply air temperature is possible resulting in a larger differential $\Delta t_c'$.

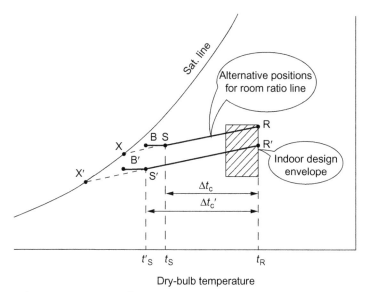

Fig. 5.5 Alternative positions of room ratio line to determine temperature differential for cooling.

Coil temperature. The minimum off-coil air temperature **B** may be limited by the cooling coil temperature. Typically, the average lowest average air temperature leaving a coil using chilled water is 8°C.

Latent Heat Gains

Latent heat gains to a space arise from the following sources:
- Infiltration of outdoor air;
- Occupants;
- Processes.

The total gain from these sources is the *design latent heal gain*, q_l.

To deal with room latent heat gain, air is supplied at a moisture content lower than that of the room air condition. The balance of the heat gain to the dehumidifying effect of the air supply is given by Eq. (2.21), i.e.,

$$q_l = \dot{m}_a h_{fg}(g_R - g_S)$$

Where there is a relatively large design latent heat gain compared with sensible heat gain, it may be necessary to use the moisture content differential to determine the design flow rate. The design *mass* air flow rate is therefore given by

$$\dot{m}_a = \frac{q_l}{h_{fg}\,\Delta g_c} \tag{5.3}$$

where $\Delta g_c = (g_R - g_S)$, the design cooling moisture content differential.

Since the room moisture content g_R is the controlled condition, the mass flow rate is determined by fixing the supply air moisture content g_S, which should be as low as possible, consistent with the performance of the air cooling coil.

Winter Heating

For winter heating, the AC system becomes an air heating system, when sensible heat losses from a space arise from the following sources:

- Heat transmission through the building fabric due to internal/external temperature differences;
- Infiltration of outdoor air.

The total loss from these sources, based on the outdoor air design dry-bulb temperature, is the *design sensible heat loss* q_s'. The heat balance of the heat loss to the heating effect of the air supply is given by:

$$q_s' = \dot{m}_a c_{pas}(t_S - t_R) \tag{5.4}$$

Therefore, if the heat loss is the design heat loss q_s', then the maximum supply air temperature, t_{Smax}, is given by:

$$t_{Smax} = \frac{q_s'}{\dot{m}_a c_{pas}} \tag{5.5}$$

If the supply air temperature is too high for satisfactory room air diffusion, either the design air mass flow rate can be increased or an alternative system of air diffusion used. This is unlikely to be a problem for a *constant air volume* (CAV) system designed for both cooling and heating, since the heat gains will be numerically larger than the heat losses. However, with variable-air-volume system terminal units, which are required to deal with a heating load, excessive supply air temperatures are likely; this problem is examined below.

Total Air Flow Rate

For CAV systems, the total air volume flow rate will normally be the sum of the flow rates to, or from, each outlet in the system.

With *variable air volume* (VAV) systems, the total air flow rate can be reduced by analysing the variation in cooling load requirements between

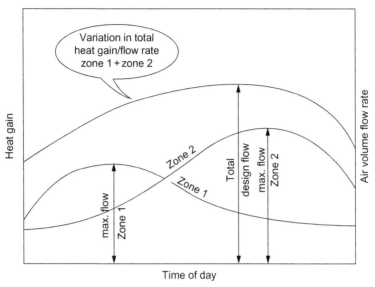

Fig. 5.6 Heat gains and air flow rates for a VAV system.

the different spaces. This is illustrated in Fig. 5.6 where the loads of two zones vary through the day and are out of phase with each other. The supply air flow rates for the individual rooms are calculated for the maximum gains to each zone, but since these design flow rates occur at different times, the maximum total of the loads for both zones may be used to determine the fan supply volume.

Variable Air Volume

To deal with the variations in sensible heat gain to an individual space using a VAV system, the flow rate is reduced, while the supply air temperature remains constant. Since the terminal unit is required to deliver a minimum flow rate to provide ventilation and adequate air diffusion, there is a limit to the reduction in gain that can be dealt with in this way. The minimum-to-maximum flow rate of a terminal unit is known as the *turndown ratio* (TDR), and for most units, this is of the order of 1:4.

Example 5.2

A room is supplied with air from a VAV terminal unit designed for cooling and heating. Determine the supply air temperature when the room is subject to the maximum heat loss for the following design data:

Maximum heat gain	10 kW
Maximum heat loss	4 kW
Room temperature	21°C
Minimum supply air temperature	12°C
Turndown ratio (TDR)	1:4

Solution

The supply air mass is given by Eq. (5.1):

$$\dot{m}_a = \frac{q_s}{c_{pas}\,\Delta t_c} = \frac{10}{1.02\,(21-12)} = 1.09\ \text{kg/s}$$

Therefore, the minimum flow rate for a TDR of 1:4, the mass flow rate, is:

$$\dot{m}_a = 0.25 \times 1.09 = 0.272\ \text{kg/s}$$

If the terminal unit continues to operate on the minimum flow rate, the maximum supply air temperature is given by Eq. (5.4):

$$t_{Smax} = \frac{q_s'}{\dot{m}_a c_{pas}} = 21 + \frac{4}{0.272 \times 1.02} = 35.4°\text{C}$$

In this example, the supply air temperature would be considered too high for successful air diffusion, and therefore, the flow rate should be increased when a maximum supply air temperature is reached. Alternatively, the air can be supplied from another outlet position to promote satisfactory room air diffusion. One such an arrangement is illustrated in Fig. 5.7. Here, the air is supplied across the ceiling for cooling, but with a rise in supply air

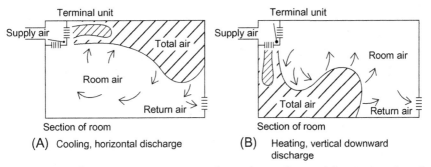

(A) Cooling, horizontal discharge   (B) Heating, vertical downward discharge

Fig. 5.7 Air diffusion with a VAV terminal unit for cooling and heating modes of operation.

temperatures to 2–3 K above room temperature, the outlet adjusted to give a vertical discharge down the wall.

MINIMUM VENTILATION RATES FOR NORMAL OCCUPANCY

Air supply for ventilation purposes is usually taken to be the supply of outdoor air into the occupied spaces. Where the building is air conditioned, this is supplied via the ductwork system, the purpose being to maintain a satisfactory indoor air quality by reducing the concentration of airborne contaminants to levels acceptable to the occupants and/or for the activities taking place within the building. A common requirement is the dilution of body odours originating from normal body processes. In the United Kingdom, there are no statutory requirements with regard to specific minimum ventilation rates. For the United Kingdom, the recommendation in the CIBSE Guide is for a minimum ventilation rate of 10 l/s per person; in the United States, a figure of 7.5 l/s is likely to become the accepted standard.

If the minimum ventilation rate is less than the total air supply rate required to meet the cooling and heating loads of the space, then the difference between the two can usually be recirculated within either the system or the room unit. Where the air supply rate exceeds the minimum ventilation rate and it is not permissible to recirculate air, the total supply air is drawn from outdoors. With VAV systems, it is often required that a minimum ventilation rate is maintained, rather than have a fixed proportion of outdoor to recirculated air. It is also worth bearing in mind that, with all-air systems designed for *free cooling*, the system will be supplied considerably more than the minimum ventilation requirement for most of the year (see Chapter 6).

AIR MOVEMENT CONTROL

Air movement control (AMC), sometimes known as room *pressurization, is* the control of air movement within a suite of rooms and within the building complex. Particular areas of interest are
- Operating room suites in hospitals;
- Isolation rooms and intensive care suites;
- Clean rooms;
- Animal laboratories;
- Corridors as escape routes;

- Smoke control zones;
- Toilets associated with offices;
- Kitchens and serveries.

The general principle is to prevent potentially contaminated air from passing from one area to a clean area by the movement of air flows. To do this, it is necessary to plan the direction of flow from a *clean zone* to a relatively *less clean* or *dirty* zone. For air to pass from one zone to another, a pressure difference must exist, but this can be quite small. What is important is that there should be sufficient air flow to minimize the risk of a reverse flow from the less clean areas to the areas requiring protection. At the same time, unless the temperatures of each space are controlled at the same level, convection currents can be promoted, which might transfer airborne contamination from a less clean area to the protected area.

A hospital operating suite is taken as an example to illustrate the principles involved. The suite of rooms (or spaces) is divided into four zones, *sterile*, *clean*, *transitional*, and *dirty*, as given in Table 5.3. This table gives the air supply rates for the dilution of airborne bacterial contaminants, which are considered sufficient, provided that reasonable mixing of room air occurs. The room pressures are given to help with the sizing and the commissioning

Table 5.3 Hierarchy of cleanliness and recommended air supply and extract rates for dilution of airborne bacterial contaminants in operating room suites

Class	Room	Nominal pressure (Pa) (i)	Air flow rate for bacterial contamination dilution	
			Flow in or supply rate	Flow out or extract rate
Sterile	Preparation room			
	(a) Layup	35	0.20	–
	(b) Sterile pack store	25	0.10	–
	Operating room	25	0.65	–
	Scrub-up bay	25	0	0 (ii)
Clean	Sterile pack bulk store	+ve	Greater of 15 ac/h or 0.15 ac/h	
	Anaesthetic room	14	–	
	Scrub-up room	14		

Continued

Table 5.3 Hierarchy of cleanliness and recommended air supply and extract rates for dilution of airborne bacterial contaminants in operating room suites—cont'd

Class	Room	Nominal pressure (Pa) (i)	Air flow rate for bacterial contamination dilution	
			Flow in or supply rate	Flow out or extract rate
Transitional	Recovery room	3	15 ac/h	15 ac/h (iii)
	Clean corridor	0	See note (iv)	7 ac/h
	General access corridor	0	See note (iv)	7 ac/h
	Changing rooms	3	7 ac/h	7 ac/h
	Plaster room	3	7 ac/h	7 ac/h
Dirty	Service corridor	0	–	(v)
	Disposal room	−5 or 0	–	0.1

(i)   Flow rates in m^3/s or air changes per hour (ac/h).
(ii) Nominal room pressures given in the table are not an essential feature of the design; they are given to facilitate setting-up pressure relief dampers, the calculation process, and the sizing of air transfer devices. The open bay scrub is considered to be part of the operating room; no specific extract is required provided air movement is satisfactory.
(iii) An air change rate of 15 ac/h is considered necessary for the control of anaesthetic gas pollution.
(iv) A supply flow rate is necessary to make up 7 ac/h after taking account of secondary air from cleaner areas.
(v)   No dilution is required; air flow rates required only for temperature control.
(vi) If two rooms are of equal cleanliness, no flow is required, and the design of the air movement will assume zero air flow. In certain cases, however, interchange is not permitted, and a protection air flow rate of 0.28 m^3/s is assumed in the design.
(Data from Department of Health, Technical Memorandum 03–01. Specialised ventilation for healthcare premises, Part A—Design and ventilation, 2007, with permission from the Department of Health)

of any pressure relief dampers considered necessary to the overall design. (It should be noted that the room pressures quoted are not essential to the design solution, rather, the design deals with the air flows through doorways.) When doors are closed, minimizing the movement of contaminated air from a dirty to a clean zone is relatively easy to achieve, but once open, the problem becomes more complex. The design of the AMC system assumes that only one door is opened at a time and that the direction and rate of air flow through that doorway should be sufficient to prevent any serious backflow of air to a cleaner area. AMC control schemes have been developed for several possible operating suite proposals. The diagrammatic layout of one scheme is shown in Fig. 5.8 [3].

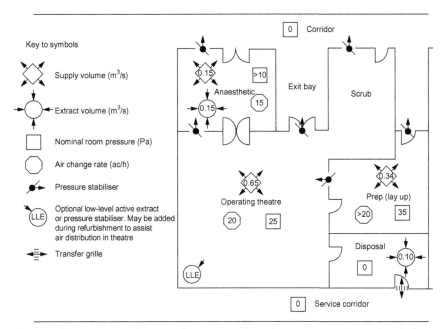

Fig. 5.8 Diagrammatic layout of an operating theatre suite. *(Reproduced from Department of Health, Technical Memorandum 03–01. Specialised ventilation for healthcare premises, Part A—Design and ventilation, 2007, with permission from the Department of Health.)*

SMOKE CONTROL ESCAPE ROUTES

In some buildings, the air conditioning and mechanical ventilation systems must be designed to assist the control of smoke in the event of fire. The aim is to keep escape corridor routes free of smoke, by maintaining a positive pressure in the corridor. To achieve this, supply and extract air flow rates are determined based on flow rates through door grilles from the corridor to individual rooms. Separate pressurization systems are provided for staircases.

SYMBOLS

c_{pas} specific heat of humid air
g moisture content
h_{fg} latent heat of evaporation
$\dot{m}$ mass flow rate of dry air
q_s sensible heat gain to air conditioned space
q_s' sensible heat loss to air conditioned space

q_1 latent heat gain to air conditioned space
t dry-bulb temperature
t_{Smax} maximum supply air temperature
$\dot{V}$ air volume flow rate
v air-specific volume
x fraction of supply air volume
γ ratio of room sensible to total heat gain
Δg moisture content differential
Δt_c temperature differential for cooling

SUBSCRIPTS

R   room air condition
S   supply air condition

ABBREVIATIONS

AMC   air movement control
MRT   mean radiant temperature
RRL   room ratio line
TDR   turndown ratio

REFERENCES

[1] A.C. Hardy, H.G. Mitchell, Building a climate: the Wallsend project, J. Inst. Heat. Vent. Eng. 38 (1970) 71.
[2] CIBSE, Guide Book B, Section B3, Chartered Institute of Building Services Engineers, 2015.
[3] Department of Health, Technical Memorandum 03–01. Specialised ventilation for healthcare premises, Part A—Design and ventilation, 2007.

CHAPTER 6

All-air Systems

In this chapter, a number of air conditioning systems are described to illustrate the principles of design. They are discussed in terms of determining the heating and cooling plant loads together with the operation of the control systems. The plant items give the process requirement whilst not specifying the heating or cooling medium.

Initially, relatively simple systems are examined in which the air supply rate remains constant; these flow rates are determined as described in the previous chapter. The systems are studied in detail so that the reader may have a clear understanding of the procedures and principles involved. Although these systems are often used in their own right, a thorough understanding of their operating principles is necessary before going on to consider more complex systems such as the variable-air-volume system.

When designing, a block diagram of the system should be built up, together with the seasonal air conditioning processes as complete cycles on a psychrometric chart. The air state points should be identified with letters (or numbers) so that the processes on the chart can be related to those on the block diagram.

SYSTEMS USING 100% OUTDOOR AIR

With a system using 100% outdoor air, a decision has already been made that, due to the ventilation requirements, no recirculation of air is allowed.

Summer Operation

For summer operation, a design procedure would be as follows: (referring to Fig. 6.1)
- Choose the room design condition **R**.
- Calculate the design sensible and latent heat gains and determine the room ratio line.
- On the psychrometric chart, plot the room condition and the room ratio line.

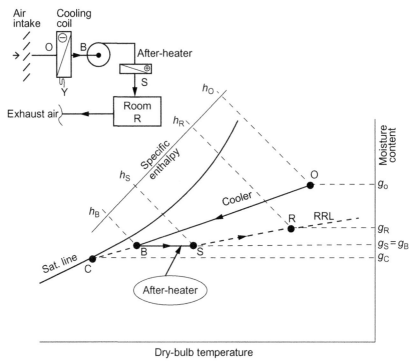

Fig. 6.1 System using 100% outdoor air, summer operation.

- Determine the supply air condition **S**, (together with the supply air mass flow rate).
- Plot the outdoor design condition **O**.

The conditions on the chart now show that it is necessary to cool and dehumidify the outdoor air from **O** to **S**, showing that a dehumidifying cooling coil is required. However, the cooling coil process line must cut the saturation line at its ADP, whereas the projection of the line joining **O** and **S** does not. To obtain the off-coil condition **B**, the contact factor of the coil is used to obtain the moisture content of **C**, knowing that the moisture content of the off-coil condition is the same as that of the supply air and that the on-coil moisture content is g_O.

The cooling process line **OBC** can then be drawn. The air at condition **B** is at a lower temperature than the supply air, and the plant therefore requires a heating coil, termed an *after heater* or *reheater*, to heat the air from **B** to **S**. In practise, the temperature rise shown as afterheater load will also include fan and duct heat gains.

With the psychrometric process now complete, the cooling coil load is obtained:

$$Q_c = \dot{m}_a(h_O - h_B) \tag{6.1}$$

The cooling coil load Q_c is made up of the following part loads:
Fresh air load is given as

$$Q_{fa} = \dot{m}_a(h_O - h_R) \tag{6.2}$$

Room load (total heat gain) is given as

$$Q_{rt} = \dot{m}_a(h_R - h_S) \tag{6.3}$$

Afterheater load is given as

$$Q_{ah} = \dot{m}_a(h_S - h_B) \tag{6.4}$$

Such a breakdown of loads illustrates that the refrigeration plant load is not equal to the total heat gain/cooling load acting on the air conditioned space; it also includes loads due to the need to reheat the air[1] and to cool and dehumidify the outdoor air. For an energy-efficient design, the engineer should match plant loads as closely as possible to the building cooling load, but this requirement does depend on the type of system chosen. In this case, once the room loads have been determined, the aim should be to minimize the ventilation and afterheater loads.

Example 6.1
A room, maintained at 21°C dry-bulb temperature and 55% rh, has heat gains of 10 kW sensible and 3 kW latent. For an air conditioning system using 100% outdoor air, determine the plant loads for the following design conditions:

Minimum supply air temperature	13°C
Outdoor design conditions	27°C dry bulb and 20°C wet bulb
Cooling coil contact factor	0.85
Specific heat of humid air	$1.02\,kJ/kg_{da}$

Solution
Refer to Fig. 6.2.
Supply air mass flow rate is given by Eq. (5.1):

---

[1] UK building regulations require that reheat is not to be used when there is cooling; the system described are used to illustrate the way a system is built up using a psychrometric chart.

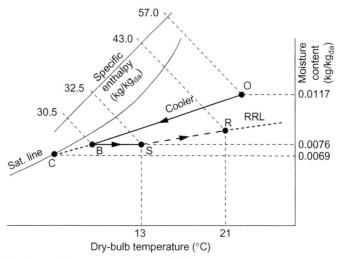

Fig. 6.2 Psychrometric processes—Example 6.1.

$$\dot{m}_a = \frac{q_s}{c_{pas}\Delta t_c} = \frac{10}{1.02\,(21-13)} = 1.23\,\text{kg/s}$$

Slope of the room ratio line is:

$$\gamma = \frac{q_s}{q_s + q_l} = \frac{10}{10 + 3} = 0.7$$

On a psychrometric chart,

- plot room condition **R** and room ratio line;
- identify supply air condition **S** on the RRL and obtain moisture content g_S;
- plot the outdoor condition **O** and obtain moisture content g_O; using the coil contact factor, determine g_C from Eq. (2.8):

$$\beta = \frac{g_A - g_B}{g_A - g_C}$$

$$0.85 = \frac{0.0117 - 0.0076}{0.0117 - g_C}$$

$$g_C = 0.0069\,\text{kg/kg}_{da}$$

- plot g_C on the saturation line to give condition **C**;
- draw the straight line **OC**; the intersection of this line with the moisture content line g_S gives the air condition **B** after the cooling coil;
- join **B** to **S**.

The psychrometric process is now complete. The cooling coil process is the line **OB**; reading the enthalpies from the chart, the cooling coil load is given by Eq. (6.1):

$$Q_c = \dot{m}_a(h_O - h_B)$$
$$= 1.23(57 - 30.5) = 32.6\,\text{kW}$$

The cooling coil load may be subdivided as follows:
Ventilation load from Eq. (6.2) is:

$$Q_{fa} = \dot{m}_a(h_O - h_R)$$
$$= 1.23(57 - 43) = 17.2\,\text{kW}$$

Room load from Eq. (6.3) is:

$$Q_{rt} = \dot{m}_a(h_R - h_S)$$
$$= 1.23(43 - 32.5) = 12.9\,\text{kW}$$

The afterheater process is the line **BS**, and the load is given by Eq. (6.4):

$$Q_{ah} = \dot{m}_a(h_S - h_B)$$
$$= 1.23(32.5 - 30.5) = 2.5\,\text{kW}$$

Note. The enthalpy values measured from the chart are not as precise as value table values; hence, the small discrepancy between the total room load and that calculated from Eq. (6.3).

Winter Operation

The operation at the winter design conditions is now considered, using the system designed for summer operation. The air conditioned space now has a sensible heat loss, q'_S, and a latent heat gain, q_l.

Since the air mass flow rate, determined for summer operation, remains constant, the supply air temperature is given by:

$$t_{Smax} = t_R + \frac{q'_S}{\dot{m}_a c_{pas}} \tag{6.5}$$

Referring to Fig. 6.3, a procedure similar to that for the summer operation is followed.

On the psychrometric chart:
- plot the room condition and the RRL,
- identify the supply air condition **S** on the RRL,
- plot the winter outdoor design condition **O**.

The conditions on the chart show that it is necessary to heat and humidify the outdoor air from condition **O** to condition **S**. Since there is already an after heater in the system, a sensible-heating process line may draw through **S**. The chart shows that the outdoor air now requires heating and humidifying from **O** to **B**. To achieve this, the system can be designed in one of the two ways:

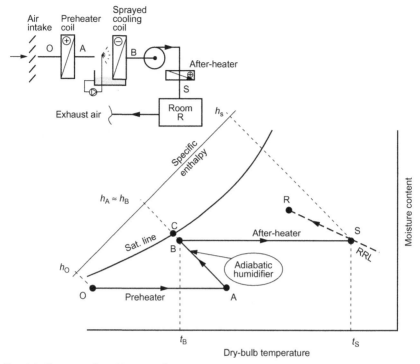

Fig. 6.3 System using 100% outdoor air, winter operation.

- Sensible heating followed by adiabatic humidification;
- Sensible heating followed by steam humidification.

SYSTEM USING ADIABATIC HUMIDIFICATION

Continue with Fig. 6.3 and assume that a sprayed cooling coil with a *humidifying* contact factor β is included in the system. On the psychrometric chart:
- draw the sensible-heating process line through **O**;
- following a line of constant wet–bulb temperature, draw **ABC** such that $\beta = \textbf{AB}/\textbf{AC}$, where:

$$\beta = \frac{g_B - g_A}{g_C - g_A}$$

and all conditions are known except g_C.

With the psychrometric process now complete, the preheater load is given by:

$$Q_{ph} = \dot{m}_a(h_A - h_O) \tag{6.6}$$

The design afterheater load is given by:

$$Q_{ah} = \dot{m}_a c_{pas}(t_{Smax} - t_B) \tag{6.7}$$

In practise, the temperature rise shown as afterheater load may include a small temperature rise from the fan and duct heat gains.

To summarize, the operation of the plant is as follows: the outdoor air is heated from condition **O** to condition **A** in order that the adiabatic process **AB** can take place. Air leaving the humidifier is then at the right moisture content to maintain the room humidity, the afterheater raising the supply air temperature to **S** to meet the room sensible load requirements.

Example 6.2

The room in Example 6.1 is maintained in winter at 19°C dry-bulb temperature and 45% rh when the design sensible heat loss is 5 kW and latent heat gain is 2 kW. With the plant now including a preheater and a sprayed cooling coil, determine the plant loads for the following design conditions:

Outdoor design condition   −4°C dry bulb, 100% sat
Sprayed coil contact factor   0.75

Solution

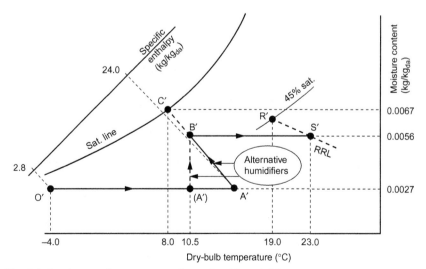

Fig. 6.4 Psychrometric processes—Examples 6.2 and 6.3.

Refer to Fig. 6.4.

From Example 6.1, the supply air mass flow rate is 1.23 kg/s.

Maximum supply air temperature is given by Eq. (6.5):

$$t_{Smax} = t_R + \frac{q'_S}{\dot{m}_a c_{pas}}$$
$$= 19 + \frac{5}{1.02 \times 1.23} = 23°C$$

Determine the room ratio line:

$$y = \frac{5}{5+2} = 0.71$$

On the psychrometric chart:

- plot room ratio line;
- identify the supply air condition **S** on the RRI and obtain moisture content, g_S;
- plot the outdoor condition **O** and obtain moisture content, g_O.

Calculate the coil contact factor:

$$\beta = \frac{g_B - g_A}{g_C - g_A}$$
$$0.75 = \frac{0.0056 - 0.0027}{g_S - 0.0027}$$
$$g_C = 0.0067 \, kg/kg_{da}$$

- Plot g_C on the saturation line to give condition **C**; this gives a wet-bulb temperature of 8°C.
- Draw the wet-bulb temperature from **C**; the intersection of this line with the moisture content line g_S gives the air condition **B** after the sprayed cooling coil.
- Join **B** to **S**; this gives the afterheater process line.
- Draw the preheater process from **O** to cut the 8°C wet-bulb temperature line to give condition **A**.

The psychrometric process is now complete.

The preheater coil load is given by Eq. (6.6):

$$Q_{ph} = \dot{m}_a(h_A - h_O) = 1.23(24 - 2.8) = 26.1 \, kW$$

The afterheater load is given by Eq. (6.7):

$$Q_{ah} = \dot{m}_a c_{pas}(t_{Smax} - t_B) = 1.23 \times 1.02(23 - 10.5) = 15.7 \, kW$$

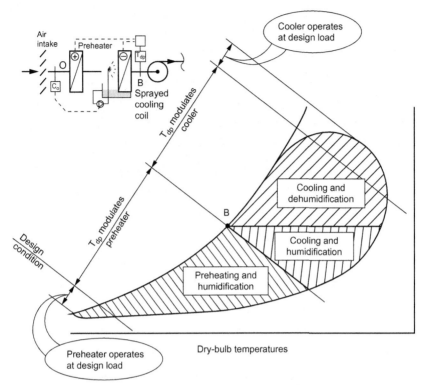

Fig. 6.5 100% outdoor air conditioning system, summary of plant operation.

Year-Round Operation of Plant

To deal with changes in the outdoor air conditions and variations in room heat gains, it is necessary to regulate the output of the heating and cooling coils. The completed system with controls, together with a summary of the operation on an outdoor condition envelope, is shown in Fig. 6.5.

The system of controls shown operates as follows: a room thermostat, T_R, controls the output of the afterheater to vary the supply air temperature in accordance with the sensible gains to the room. The dry-bulb thermostat T_{dp} controls the output of the cooling coil and preheater in sequence to maintain the supply air moisture content and hence room humidity. Since the air leaving the cooler is close to the saturation line (and the moisture content of this air condition is the supply air moisture content), this method of controlling humidity is known as *dew-point* control.

When this method of control is used, variations in room humidity can occur from the following:

- Contact factor of the sprayed cooling coil;
- Proportional band of dew-point controller;
- Variations in room latent gain;
- Average air conditions in the duct not sensed by the thermostat T_{dp} under varying loads.

For most systems, variations in humidity will be acceptable in that the room condition can be maintained within the indoor condition envelope. If close control of humidity is required, then T_{dp} should be reset by a humidistat mounted either in the room or in the extract air duct.

The system will operate satisfactorily with the humidifier in operation all year round since, even with water sprayed onto the cooling coil, dehumidification will occur. However, in order to save on pumping energy costs, the humidifier can be switched off by sensor C_O when humidification is not required. Referring to Fig. 6.5, the ideal control is one in which this sensor measures either the moisture content or dew-point temperature of condition **B**. Since these ideal sensors are relatively expensive compared with the energy saved, a less satisfactory but cheaper method of switching off

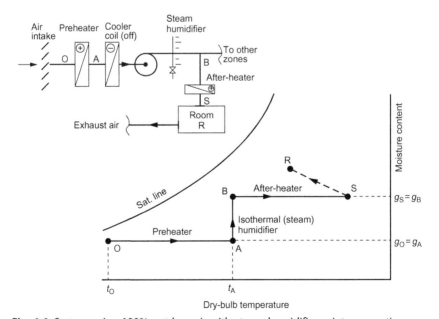

Fig. 6.6 System using 100% outdoor air with steam humidifier, winter operation.

the pump with C_O as a dry-bulb thermostat could be considered though it is more usual to exclude it from the system.

SYSTEM USING STEAM HUMIDIFICATION

The second route to achieving the required supply air moisture content is by using a steam humidifier. Refer to Fig. 6.6.

Assume that the steam humidifier is placed after the fan. The preheater must raise the air temperature so that the required amount of steam can be added below the saturation line. On the psychrometric chart,

- draw the sensible-heating process line through **O**.
- draw the steam humidifier process line to give condition **B** such that $g_B = g_S$.

With the psychrometric process now complete, the preheater load is given by:

$$Q_{ph} = \dot{m}_a c_{pas}(t_A - t_O) \tag{6.8}$$

the *steam humidifier* load is given by:

$$Q_{st} = \dot{m}_a(h_B - h_{BA}) \tag{6.9}$$

the steam supplied is given by:

$$\dot{m}_s = \dot{m}_a(h_S - h_B) \tag{6.10}$$

The design afterheater load is given by Eq. (6.7).

Summary of Plant Operation

The outdoor air is heated from condition **O** to condition **A** in order that the steam humidification **AB** can take place. Air leaving the humidifier is at the correct moisture content to maintain the room humidity, the afterheater raising the air to the supply temperature **S** to meet the room sensible load requirements.

Example 6.3

By using relevant data from Example 6.2, determine the plant loads using a steam humidifier in place of an adiabatic humidifier. The temperature after the preheater is 10°C.

Solution

Refer to Fig. 6.6.

On the psychrometric chart, the room ratio line, supply, and outdoor conditions have been plotted, using data from Example 6.2.

Supply air mass flow rate is 1.23 kg/s.

Moisture contents are given by:

Outdoor air g_O: 0.0027 kg/kg$_{da}$

Supply air g_S: 0.0056 kg/kg$_{da}$

Preheat load is given by Eq. (6.8):

$$Q_{pt} = \dot{m}_a(h_A - h_O)$$
$$= 1.23 \times 1.02\,[10.5 - (-4)] = 18.2\,\text{kW}$$

Steam humidifier load is given by Eq. (6.9):

$$Q_{st} = \dot{m}_a(h_B - h_A)$$
$$= 1.23 \times 1.02(24.5 - 17.5) = 8.6\,\text{kW}$$

Steam supply is given by Eq. (6.10):

$$\dot{m}_s = \dot{m}_a(h_S - h_B)$$
$$= 1.23(0.0056 - 0.0027) = 0.00357\,\text{kg/s}$$

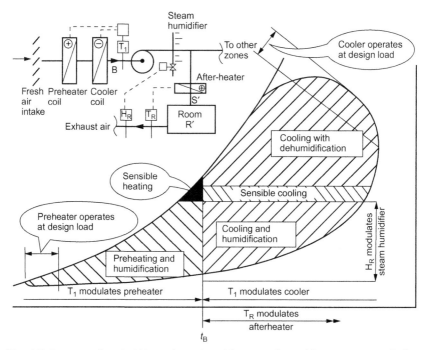

Fig. 6.7 System using 100% outdoor air, with steam humidifier, summary of plant operation.

Year-Round Operation of Plant

As with the system using an adiabatic humidifier, it is necessary to regulate the output of the heaters, humidifier, and cooling coil. The completed system with controls, together with a summary of its operation, is summarized on an outdoor condition envelope (see Fig. 6.7).

The system of controls operates as follows: the thermostat T_R controls the output of the afterheater to vary the supply air temperature in accordance with the sensible gains to the room. The dry-bulb thermostat T_1 controls the output of preheater and cooling coil in sequence. This thermostat is now only partially dew-point control; the cooling coil is dehumidifying when the dew-point of the outdoor air is above the controlled condition between the coil and the fan. For the remainder of the year, when the outdoor air dry-bulb temperature is higher than that of the controlled condition t_B, the coil provides only sensible cooling, whilst the humidistat regulates the output of the steam humidifier to maintain the room humidity.

SYSTEMS USING RECIRCULATED AIR

By recirculating air from the air conditioned space(s), the cooling coil load can be reduced and the preheater either omitted or considerably reduced in capacity, compared with using 100% outdoor air. This leads to savings in both capital and energy consumption costs.

The minimum quantity of outdoor air is determined by the fresh air requirements of the room.

SYSTEM USING A FIXED PERCENTAGE OF OUTDOOR AIR

Summer Operation

Consider a fixed percentage of outdoor air, determined by the fresh air requirements. Using a design procedure similar to that described for the system using 100% outdoor air, the block diagram and the psychrometric process for summer operation are developed in Fig. 6.8. (Note that the humidifier is required for winter operation; where there is no humidifying requirement, these will be omitted.)

On the psychrometric chart,
- plot the room condition **R** and the room ratio line, RRL;
- identify the supply air condition **S** on the RRL;

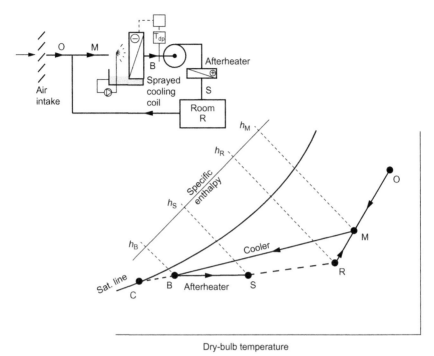

Fig. 6.8 Recirculation air conditioning system, summer operation.

- plot the outdoor air design condition **O**;
- join **O** and **R** and obtain the mixed air condition **M**;
- the cooling coil process line **MB** is drawn for the coil contact factor **MB/MC**;
- the process **BS** is an afterheater load (including fan and duct heat gains). With the psychrometric process now complete, the cooling coil load is given by:

$$Q_c = \dot{m}_a (h_M - h_B) \qquad (6.11)$$

As before, this load may be broken down into *room*, *fresh air*, and *afterheater* loads. The room load (total heat gain) and afterheat loads are the same as that for the 100% outdoor system, using Eqs (6.3), (6.4). The fresh air load is given by:

$$Q_{fa} = \dot{m}_a (h_M - h_R) \qquad (6.12)$$

Therefore, compared with the system using 100% outdoor air, the cooling coil load is reduced by:

$$Q_{diff} = \dot{m}_a(h_O - h_M)$$

Example 6.4

The air conditioning system in Example 6.1, with the same room heat gains and design conditions, is to be modified so that of the total supply air 75% is recirculated and 25% is fresh air. Calculate the new cooling coil plant load.

Solution
Refer to Fig. 6.9.
From Example 6.1, the total air flow rate is 1.23 kg/s, and room ratio line is given by:

$$y = 0.77.$$

Using the psychrometric chart:
- plot room condition **R** and the room ratio line. Obtain the room moisture content:

$$g_R = 0.0087\,kg/kg_{da}$$

- identify the supply air condition **S** on the RRL, moisture content:

$$g_S = 0.0076\,kg/kg_{da}$$

- plot the outdoor condition **O** with a moisture content

$$g_O = 0.00117\,kg/kg_{da}$$

- join **O** and **R**

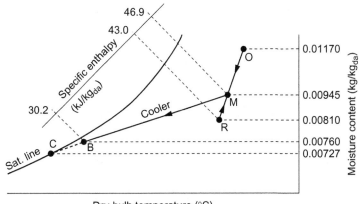

Fig. 6.9 Psychrometric processes—Example 6.4.

Determine mixed air moisture content from air mixing ratio. Using Eq. (2.2):

$$g_M = xg_A + (1-x)g_B$$
$$= 0.25 \times 0.0117 + (1-0.25)\,0.0087$$
$$= 0.00945\,kg/kg_{da}$$

Coil contact factor is given by:

$$\beta = \frac{g_M - g_S}{g_M - g_C}$$

$$0.85 = \frac{0.00945 - 0.0076}{0.00945 - g_C}$$

$$g_C = 0.0073\,kg/kg_{da}$$

- plot g_C on the saturation line to give condition **C**,
- draw the straight line **BC**,
- the intersection of this line with the moisture content line g_C gives the air condition **B** after the cooling coil.

The cooling coil load is given by Eq. (6.11):

$$Q_c = \dot{m}_a(h_M - h_B)$$
$$= 1.23\,(46.7 - 30.2) = 20.3\,kW$$

The ventilation load of the cooling coil is obtained from Eq. (6.12):

$$Q_{fa} = \dot{m}_a(h_M - h_R) = 1.23\,(46.7 - 43) = 4.6\,kW$$

Compared with the system using 100% outdoor air in Example 6.1, the cooling load has been reduced from 32.6 kW to 20.3 kW. This reduction has been achieved, in the main, by the reduced ventilation load, from 17.2 to 4.6 kW.

Winter Operation

Consider first the operation at the winter design condition for a system using a relatively small ratio of outdoor to recirculated air. Again, following the design procedure and referring to Fig. 6.10,
- plot the room condition **R** and the room ratio line;
- identify the supply air condition **S** on the RRL;
- plot the winter outdoor design condition, **O**;

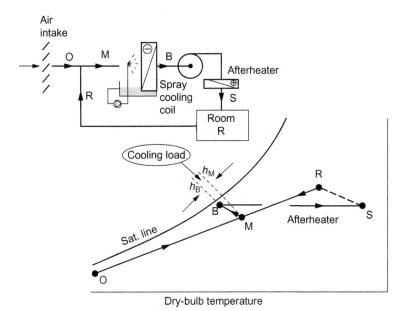

Fig. 6.10 Recirculation air conditioning system, winter operation.

- join 0 and R to obtain the mixed air condition **M**;
- draw the afterheater process through **S**;
- the process through the sprayed cooling coil is **MB**.

The psychrometric cycle is now complete. This cycle shows the enthalpy of the mixed air condition as greater than that of the controlled off-coil condition **B**. Cooling is therefore required in winter, a system operation, which is not efficient in energy use. To avoid the use of the refrigeration plant in winter, the air conditioning system may be modified to increase the percentage of outdoor to recirculated air by using modulating dampers, in a *free cooling* system, as described later.

The afterheater load is calculated from Eq. (6.4).

Consider next the operation at the winter design condition for a system using a relatively large ratio of outdoor to recirculated air. The enthalpy of the mixed air condition would be lower than the controlled off-coil condition **B**. A preheater is therefore required; the load of which is given by:

$$Q_{ph} = \dot{m}_a(h_B - h_M) \tag{6.13}$$

An alternative to placing the preheater in the mixed air stream is to place it in the duct carrying the outdoor air. Though the load of the preheater is the same as that given by Eq. (6.13), it may be more conveniently expressed by:

$$Q_{ph} = x\dot{m}_a(h_B - h_M) \tag{6.14}$$

where x = fraction of outdoor air in the total air supply $\dot{m}_a$.

The advantages of placing the preheater in the fresh air duct are:

- smaller duct size.
- improved heat transfer between primary heat supply and outdoor air. This is particularly important if the preheater is a heat recovery unit using exhaust room air (see Chapter 12).

RECIRCULATION SYSTEM USING MODULATING DAMPERS

Free Cooling System

With the fixed percentage recirculation system in which a relatively small percentage of outdoor air is used, it was seen that in winter the cooling coil

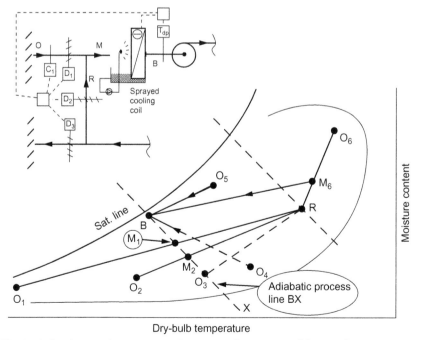

Dry-bulb temperature

Fig. 6.11 Psychrometric processes of year-round operation of free cooling system.

was required to operate at low outdoor enthalpies. A more efficient method would be to use the cold outdoor air instead of the refrigeration plant to achieve the required off-coil condition. Referring to Fig. 6.11, by increasing the outdoor air and reducing the recirculated air flow rates, the mixed air condition is such that it coincides with the adiabatic process of the humidifier. To achieve the required changes of flow rates, modulating control dampers are included in the system.

Refer to the block diagram in Fig. 6.11. Damper D_1 is fitted in the outdoor air duct and damper D_2 in the recirculation air duct. Damper D_3 is incorporated to ensure that the total flow rate in the system remains approximately constant. These dampers are adjusted by the dew-point thermostat T_{dp} to maintain a near-constant air condition **B** between the cooling coil and the fan. To illustrate how the system works through the year, a number of processes are shown on Fig. 6.11 using representative conditions within the outdoor air condition envelope, progressively increasing in enthalpy.

- *Outdoor air enthalpy* O_1 (which could be the winter design condition). The mixed air condition is M_1 such that it has the same wet-bulb temperature as **B**, taking this to be the adiabatic process through the humidifier, **BX**.
- *Outdoor air enthalpy* O_2. Damper D_1 will open to increase the outdoor air flow rate, and damper D_2 will close to reduce the recirculated air flow rate so that the mixed air condition M_2 remains on the line **BX**. Damper D_3 will open to increase the exhaust air flow rate.
- *Outdoor air enthalpy* O_3. Damper D_1 will now be fully opened, and damper D_2 is fully shut. The outdoor air is now the air condition on the line **BX**, with the system operating on 100% outdoor air.
- *Outdoor air enthalpy* O_4. The system will continue to use 100% outdoor air; the cooling coil now comes into operation, with the water sprays providing humidification.
- *Outdoor air enthalpy* O_5. The system continues to use 100% outdoor air with the cooling coil dehumidifying. The washer pump could now be switched off.
- *Outdoor air enthalpy* O_6. The system now operates on minimum outdoor air and maximum recirculation, the original summer design requirement. The dampers must therefore be changed over from 100% to minimum outdoor air operation.

The damper changeover is achieved by having a sensor C_1 located in the outdoor air duct. For energy efficiency, C_1 should measure enthalpy with a set point at the room condition; since the room condition may be subject

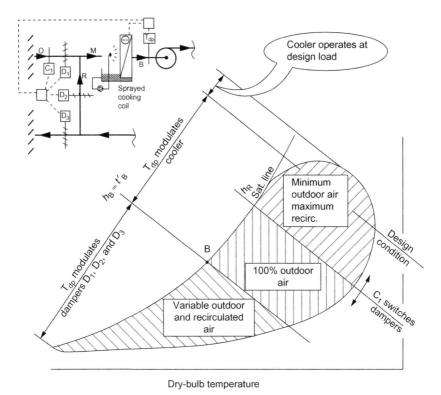

Fig. 6.12 Free cooling system with adiabatic humidifier, summary of plant operation with C_1 as an enthalpy sensor.

to small variations (within the indoor condition envelope), another sensor C_2 in the recirculation air duct may be used to reset C_1. This method of damper changeover is known as an *enthalpy control system*.

The use of modulating dampers to vary the proportion of outdoor and recirculation is termed *a free cooling system*. This is because relatively cold outdoor air is used to provide cooling (according to the needs of the system), instead of using the refrigeration plant, year round.

Year-Round Operation

The year-round operation of this system is summarized on Fig. 6.12. Depending on outdoor conditions, the cooling coil will be required for sensible cooling and dehumidification, the water sprays for humidification. The off–coil condition **B** is maintained by the dew-point thermostat T_{dp}. The coil and the control dampers D_1, D_2, and D_3 are operated sequentially by the T_{dp}. At outdoor air enthalpies below that corresponding to air condition **B**, the

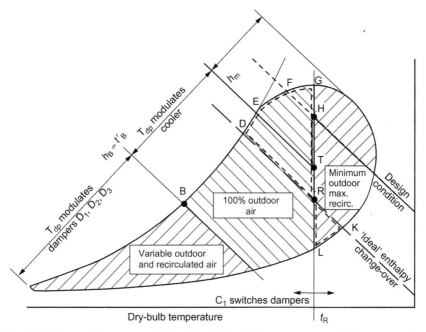

Fig. 6.13 Free cooling system with adiabatic humidifier—summary of plant operation with C_1 as a dry-bulb thermostat.

ratio of fresh to recirculated air flow rates is varied so that the mixed air condition **M** coincides with the adiabatic humidification process line **MB**. The sensor C_1, set at the room air enthalpy, effects the damper changeover.

It was shown previously that a preheater is required when the minimum ventilation air flow rate is relatively large. For this recirculation system, modulating dampers can also be incorporated to provide free cooling. In this case, the cooling coil, dampers, and preheater are controlled in sequence, and the sensor C_1 again effects the damper changeover.

Sensor C_1 as Dry-bulb Thermostat

Usually, an enthalpy signal is generated from a combination of measurements from dry bulb and relative humidity sensors. At the present time, such an enthalpy controller, though theoretically the most energy-efficient for system operation, is expensive compared with a dry-bulb thermostat. Also relative humidity sensors have been found to deteriorate in performance. Therefore, to reduce the capital cost and to obtain reliable performance, it is more usual to use the alternative of a dry-bulb thermostat at C_1. The operation of such a system is summarized in Fig. 6.13.

Some of the features of this figure are as follows:

- As before the cooling, coil and dampers are controlled in sequence by T_{dp}.
- The sensor C_1 effects the damper changeover at the dry-bulb temperature given by line GL (in this case, set at the room temperature):
  - On rise in outdoor temperature, dampers go to minimum outdoor air operation;
  - On fall in outdoor temperature, dampers go to 100% outdoor air operation.
- The ideal changeover line is the enthalpy line **DRK**. The operation of the system in diagram areas **WRLN and GRKP** is required for efficient operation. Areas **DGR and KLR** represent the time when the system is operating inefficiently.
- The cooling coil design load is based on the enthalpy of the mixed air condition, shown by line **EJ**. Therefore, when the outdoor air is in the area **EGJ**, the cooling coil load will be insufficient to maintain the controlled condition **B**. The area **EFHJ** therefore represents the additional frequency at which the coil operates above the design load.
- The set point of C_1 is shown as the room temperature; this may not be the most energy-efficient setting. The objective is to minimize the additional energy consumption in the areas **DGR and KLR**. Energy calculation methods are given in Chapter 17.
- To avoid a conflict between the operation of T_{dp} modulating the dampers and C_1 switching them over, line **BN** should not cross **GL**. The set points can be obtained by examining their relative positions on an outdoor air design envelope. If there is a danger of crossover, one of the controls must be given priority.

An alternative to resetting the dampers through sensor C_1 is to use a switch on the cooling coil control valve, set say at half the valve stem travel.

Recirculation System With No Humidifier

Many of the operating principles described above apply to recirculation systems, which do not include an adiabatic humidifier.

A diagram of the central plant is shown in Fig. 6.14.

In summer, the psychrometric cycle of this system is identical to that shown in Fig. 6.1, with the cooling load determined from Eq. (6.11). During

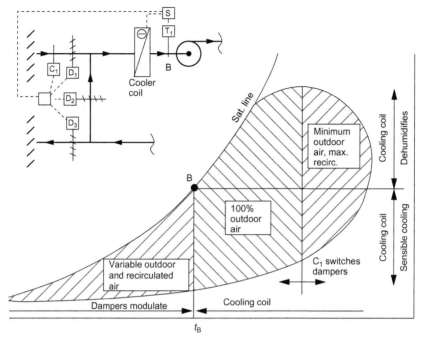

Fig. 6.14 Free cooling system with no humidifier, summary of plant operation with C_1 as a dry-bulb thermostat.

winter, when the cooling coil is switched off, the dampers are modulated by the thermostat T_1 so that the mixed air dry-bulb temperature is t_B. Since there is no humidifier in the system, the room humidity will vary with the latent heat gains. With no latent gains, the room moisture content will equal that of the outdoor air.

As before, the cooling coil and dampers are controlled in sequence by a dry-bulb thermostat T_1. The system is only partially dew-point control when the on–coil air dew-point temperature is higher than the coil ADP; during the remainder of the year, the system is under temperature control.

The controller C_1 (either an enthalpy or a dry-bulb temperature sensor) switches over the dampers as described previously; on this figure, a dry-bulb thermostat is indicated. In this case, overcooling and corresponding reheat loads will occur when the cooling coil is operating with the outdoor dry-bulb temperature above that maintained at t_B. To improve the efficiency of the system, the temperature at thermostat T_1 can be reset by an outdoor thermostat according to a schedule determined by an analysis of the room heat gains.

A steam humidifier can be incorporated in the system to maintain space humidity, similar to that described previously for the 100% outdoor system.

ZONES

Most air conditioning systems are required to deal with more than one room in a building and to ensure that design conditions are maintained throughout the building will usually have to be divided into zones. Zoning can be dealt within a number of ways, depending on the system. The number of zones will affect the number of plant items, ducting, piping, and controls. The more zones there are, the greater the capital costs of the installation, and this has to be set against increased efficiency and greater user satisfaction. It may be economical to use separate plants or air handling units to avoid excessively long runs of ducts but, even so, each of these plants may have to supply more than one room. The question arises as to which rooms should be supplied from the same zone heater or terminal unit. The decision will depend on the tolerance allowed on the optimum room air condition and the effect of loads and ventilation requirements.

Heat gains and losses vary according to the time of day and year. If these heat loads are out of phase with one another, the variations between rooms might be such as to warrant separate treatment. Thus, usually buildings with different aspects will require zones differentiated by the orientation since the solar heat gains on each of the building faces will be out of phase.

Effect of Ventilation Requirements

If the supply air rates are based on either the ventilation or air movement control requirements, the ratio of heat gain to air flow rate may differ. The difference between the ratios will indicate whether or not the spaces should be treated separately. This is illustrated by the following example:

Example 6.5

Compare the supply air temperatures for two rooms with design air volume flow rates and heat gains listed below; both rooms are maintained at 21°C. The specific volume of the supply air is $0.82\,\text{m}^3/\text{kg}$.

Room	Supply air flow rate $\dot{V}$ (m^3/s)	Heat gain Q_s (kW)
1	1.8	12
2	1.1	4

Solution

To obtain the supply air temperature, use Eq. (2.20):

$$q_s = \dot{m}_a c_{pas}(t_R - t_S)$$

Room 1

$$12 = (1.5/0.82) \times 1.02(21 - t_S)$$
$$t_S = 14.6°C$$

Room 2

$$4 = (1.1/0.82) \times 1.02\,(21 - t_S)$$
$$t_S = 18°C$$

With one heater in the supply air to both rooms, the room temperature difference would be 3.5 K. If this difference were considered unacceptable, then separate zone heaters would be required.

When more than one room is air conditioned from a central plant, the relative humidity in each of the spaces will be an average condition. If more than one room requires close control of humidity, either separate plants are required for each room can be provided with its own (zone) steam humidifier.

VARIABLE AIR VOLUME SYSTEM

Compared with constant flow rate/all air systems, variable-air-volume (VAV) systems have smaller central plant sizes and potential for energy saving. The reasons for this are that the air supplied by the fan deals with the total *instantaneous* heat gain, and there is no reheating of the air as in zoned reheat systems.

The simplest system is particularly suited for use in the following:

- Internal areas of deep plan buildings, when no heating is required in winter,
- For perimeter zones when cooling is being added to an existing building in which there is an independent heating circuit.

Possible disadvantages may arise from:

- lack of humidity control,
- insufficient supply of fresh air at low flow rates,
- unsatisfactory air distribution at low flow rates,
- unbalanced extract air flow rates.

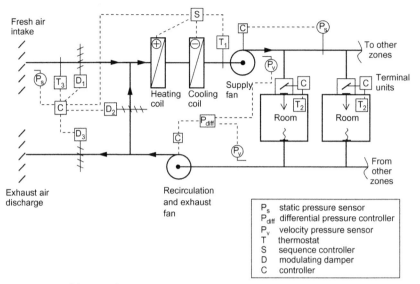

Fig. 6.15 Variable air volume system.

Air supply flow rates for this system, together with aspects of air diffusion, have been discussed in Chapter 5. Fan selection for VAV systems is described in Chapter 15.

System Arrangement and Operation

A diagram of a typical VAV system is shown in Fig. 6.15. The central plant consists of:

- control dampers,
- preheater,
- cooling coil,
- supply fan,
- terminal units.

The central plant operation will be similar to that described above, with the cooling coil, dampers, and preheater controlled in sequence by the thermostat T_1. Damper changeover is achieved through T_3 sensing the outdoor air condition.

When required to maintain the minimum ventilation rate, P_{v3} operates the dampers, if necessary overriding thermostat T_1. P_{v3} may be either a velocity, static pressure, or flow rate sensor.

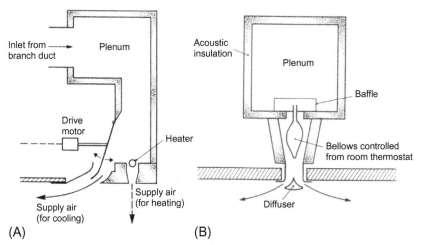

Fig. 6.16 VAV terminal units. (A) variable area air terminal device with reheat (B) throttling air terminal device, variable velocity. *(Redrawn from Fig. B3.27 of the CIBSE Guide, with permission from the Chartered Institution of Building Services Engineers.)*

A preheater is required for initial start-up and for normal operation when the minimum ventilation requirements determine that the system operates with relatively large flow rates of outside air.

The flow rates from the terminal units (TU) are regulated through the room thermostats T_2. It is usual for the TUs to operate with a (minimum) duct static pressure; as they throttle down to deal with reduced room cooling loads, the pressure in the supply air duct will rise. Rather than allow the TUs to operate against this increase in pressure, and at the same time, to encourage a reduction in fan power consumption, the supply fan flow rate is adjusted, through the static pressure sensor P_s, to maintain the duct base pressure.

The duty of the recirculation/extract fan is matched to the supply fan to maintain a balanced supply and extract within the building. A differential pressure sensor P_{diff} varies the extract fan flow rate by comparing relevant air velocities (dependent on the duct sizes) with sensors, P_{v1} and P_{v2}, in the main supply and extract ducts.

Terminal Units

The principle of varying the supply air volume flow rate to deal with changes of heat load has been described above. There are various types of terminal

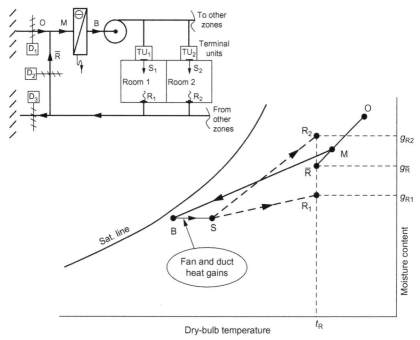

Fig. 6.17 Psychrometric process for VAV system, summer operation.

units; two of the most common are illustrated in Fig. 6.16. In (A), the air flow rate is throttled by varying the outlet area; the device shown here includes a reheater to deal with room load when the terminal is operating at the minimum turndown ratio. In (B), the flow rate is reduced by inflating a bellow device within the unit; the outlet velocity varies as the flow rate changes, and the throw across the ceiling is maintained by the Coanda effect of the air as it leaves the aerodynamically designed diffuser. The terminal units will include an acoustic lining to reduce air noise to acceptable levels.

Psychrometrics

Summer Operation

The psychrometric cycle for summer operation is shown in Fig. 6.17, for two representative rooms. The temperature at **B** is maintained by thermostat T_1. Fan and duct heat gains produce the temperature rise **BS** to give the single supply air condition, **S**. Room 1 has relatively small sensible-to-latent heat gains with room ratio line **SR₁**, whereas Room 2 has relatively large sensible-to-latent heat gains with room ratio line **SR₂**. From this, it is seen that humidity will vary between the individual rooms. The return air **R** is the average value of all room conditions.

The design cooling coil load is given by Eq. (6.11); in this equation, the supply air mass is determined by analysing the maximum instantaneous heat gain as outlined in Chapter 5.

Room humidity will also vary within a single space as the sensible gains vary; such variations are demonstrated by the following example.

Example 6.6

In a VAV system, the supply air is maintained at a dry-bulb temperature of 13°C and a moisture content of 0.008 kg/kg$_{da}$. For the following design criteria, determine the rise in room relative humidity when the sensible heat gain in one of the air conditioned rooms falls to 10 kW, the latent gain remaining constant.

Design heat gains
Sensible q_s 40 kW
Latent q_s 5 kW
Supply-to-room air temperature differential, Δt_c 8 K
Terminal unit turn down ratio   1:4
Humid specific heat, c_{pas} 1.02 kJ/kg K
Latent heat of evaporation, h_{fg} 2450 kJ/kg

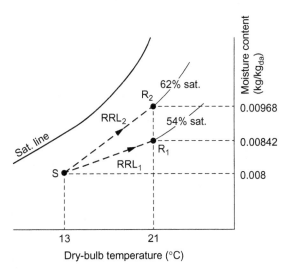

Fig. 6.18 Psychrometrics for VAV system—Example 6.6.

Solution

Refer to Fig. 6.18.

Design mass flow rate is given by Eq. (5.1):

$$\dot{m}_a = \frac{q_s}{C_{pas}\Delta t_c}$$
$$= \frac{40}{1.02 \times 8} = 4.9\,\text{kg/s}$$

At design load

The rise in moisture content from supply-to-room air is given by Eq. (2.21):

$$q_s = \dot{m}_a h_{fg}(g_R - g_S)$$

$$\Delta g = (g_R - g_S) = \frac{5}{4.9 \times 2450} = 0.00042\,\text{kg/kg}_{da}$$

The room moisture content $= g_S + \Delta g$

$$= 0.008 + 0.00042 = 0.00842\,\text{kg/kg}_{da}$$

At a room temperature of 21°C, this moisture content gives a relative humidity of 54%.

At reduced room load

Since the ratio of reduced room load to design load is 1:4, the terminal unit(s) will also operate at the turndown ratio of 1:4. The latent heat gain remains constant, and the rise in moisture content $\Delta g'$ will therefore be:

$$\Delta g' = 4 \times 0.00042 = 0.00168\,\text{kg/kg}_{da}$$

$$\therefore \text{ the room moisture content} = 0.008 + \Delta g'$$

$$g_R = 0.00 + 0.00168 = 0.00968\,\text{kg/kg}_{da}$$

At a room temperature of 21°C, this gives a relative humidity of 62%, an increase of 8% above the design load condition.

Winter Operation

A psychrometric cycle for winter operation is shown in Fig. 6.18; the terminal units incorporate heaters for winter heating. The cooling coil is switched off, and the temperature at **B** is the mixed air condition, controlled by thermostat T_1 through the modulating dampers. Both rooms have similar sensible heat gains, but Room 1 has a relatively large latent heat gain to give room ratio line **SR**$_1$. Room 2 has a negligible latent heat gain to give room ratio line **SR**$_2$. As with summer operation, it is seen that humidity will vary

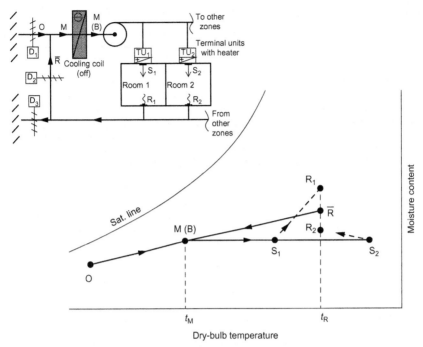

Fig. 6.19 Psychrometric process for VAV system, winter operation.

in the different rooms, and the return air is the average value of all room conditions (Fig. 6.19).

Ventilation Rate

Consideration should be paid to the supply rate of fresh air if ventilation standards are to be maintained. It has already been shown that for most of the year a system designed for free cooling will operate on considerably more than the minimum percentage of outdoor air. However, as the total instantaneous heat gain reduces, the fresh air supply rate will also fall in proportion to the reduction in total supply flow rate. Then, if the air conditioned spaces remain fully occupied, the minimum ventilation requirements will not be met. One solution to this problem is to maintain, through the controller P_1 acting on the dampers (as described above), the minimum flow rate at the fresh air intake. But even if this facility is incorporated into the system, ventilation rates may still be inadequate as is illustrated in the following example:

Example 6.7

Four zones in a building, air conditioned with a VAV 'system, have design flow rates as given in the table below. The total flow of 15 m³/s is based on an analysis of the total instantaneous cooling load of the building.

The zone ventilation rates are based on maximum occupancy; the total of these determines the minimum flow rate in the central plant of 1.9 m³/s, but the designer arbitrarily increases this to 2.25 m³/s to provide 15% of the total flow rate. Compare the ventilation rates of the four zones:

(a) Ventilation flow rate as 15% of total supply flow rate

(b) Ventilation flow rate maintained at design flow rate of 2.25 m³/s

Solution

With the system in use, the zone cooling loads are out of phase, and the zone flow rates are reduced as shown in the table, giving a total supply flow rate of 11.2 m³/s.

(a) Ventilation flow rate as 15% of total

All zone supply rates include 15% fresh air. If the zones have the expected occupancy, then the ventilation air for zones 2 and 4 fall below the required level.

(b) Ventilation flow rate maintained at design flow rate of 2.25 m³/s

The percentage of total flow rate is now $(2.25/11.2) \times 100 = 20\%$. All zone supply rates include 20% fresh air. If the zones have design occupancy, then the ventilation air for zone 2 is now satisfactory, but zone 4 still falls below the required level.

System condition	Zone				Total flow rate
	1	2	3	4	
At design load					
Maximum supply flow rate	5.50	4.00	6.00	4.00	15.00
Minimum supply flow rate	0.55	0.40	0.60	0.40	2.25
At part load					
Reduced supply flow rate	4.00	2.00	4.00	1.20	11.20
Ventilation rate—15%	0.60	0.30	0.60	0.18	1.68
Ventilation rate—20%	0.80	0.40	0.80	0.24	2.24

DUAL DUCT SYSTEM

Both warm and cold air were supplied from a central plant to each air conditioned space, combined in mixing boxes according to the cooling/heating needs of the space. The dual duct system was used where:

• relatively large ventilation/air change rates are required,

- automatic control of individual spaces or varieties of room loads
- preclude zoning,
- piped (wet) services cannot be used,
- control of humidity is not required.

Compared with other systems, this system was relatively energy-intensive, appropriate to the declining energy costs in the 1960s.

SYMBOLS

c_{pas}	specific heat of humid air
$\dot{m}_a$	mass flow rate of dry air
$\dot{m}_s$	mass flow rate of steam
Q_c	cooling coil load
Q_{ah}	afterheater load, part load of cooling coil
Q_{fa}	fresh air load, part load of cooling coil
Q_{ph}	preheater load
Q_{rt}	room total load, part load of cooling coil
Q_s	steam humidifier load
Q_h	heating coil load
q_s	sensible heat gain to air conditioned space
q_s'	sensible heat loss from air of conditioned space
q_l	latent heat gain to air conditioned space
$\dot{V}$	air volume flow rate
β	contact factor
g	moisture content
h	specific enthalpy of humid air
t	dry-bulb temperature
x	fraction of total supply air
y	ratio of room sensible to total heat gain

SUBSCRIPTS (FOR TEMPERATURE, MOISTURE CONTENT, AND ENTHALPY)

C	air condition on saturation line
M	mixed air condition
O	outdoor air condition
R	room condition
S	supply air condition

ABBREVIATION

RRL   room ratio line

CHAPTER 7

Unitary Systems

Where high air-change rates are not required (e.g., for air movement control), unitary systems will usually be used; mostly, they are used in medium-to-large buildings such as office blocks and hotels. The design and selection of a system will be based on a detailed analysis of the heat gains and losses together with the establishment of flow diagrams and heat balances. For certain periods of the year, different parts of a building may require heating and cooling simultaneously; for overall system efficiency, it is desirable to take account of load diversity patterns. The final selection of a system will depend on many factors including an analysis of both the capital and energy costs of the system.

The systems described in this chapter are as follows:
- Air and water;
- Unitary heat pumps;
- Variable refrigerant flow;
- Self-contained units.

AIR AND WATER SYSTEMS

Compared with air, the heat-carrying capacity of water is very high; therefore, to reduce the space taken up in the building by the fluid flow networks, heated and cooled water is generated centrally and piped through the building to room units, with fresh air for ventilation supplied from a central plant (or occasionally brought in at the unit through the external wall). There can be considerable flexibility in the overall design and the way the units are arranged within the building.

Air and water systems unitary systems include:
- Induction unit;
- Fan coil.

INDUCTION UNIT SYSTEMS

The principle of the induction unit is as follows: a *primary air* supply is generated in a central plant room; this air is sufficient to meet the ventilation

Air Conditioning System Design
http://dx.doi.org/10.1016/B978-0-08-101123-2.00007-8

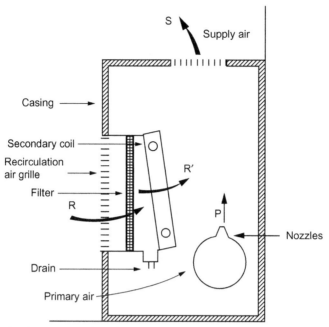

Fig. 7.1 Induction unit, single coil.

requirements of the treated space and is supplied at a relatively high pressure to a header duct in the terminal unit. Referring to Fig. 7.1, the primary air **P** is discharged through a set of nozzles, and this induces *secondary* (recirculated) air **R** from the room into the unit and over the secondary coil. The secondary air is then cooled (and/or heated, depending on the design) to **R'** mixed with the primary air to give the supply air condition **S** from the unit to the room.

The induction ratio I_r is the ratio of the induced volume air flow rate to the primary air, i.e.,

$$I_r = \frac{\dot{V}_r}{\dot{V}_p} \qquad (7.1)$$

Without significant error, the induction ratio can also be expressed in terms of the mass flow rates, i.e.,

$$I_r = \frac{\dot{m}_r}{\dot{m}_p} \qquad (7.2)$$

Induction units have a typical inlet pressure to the nozzle of 150–250 Pa; dampers are provided to regulate the primary air supply and nozzle pressure. Room temperatures are maintained by a thermostat operating two-port control valves; alternatively, a damper may be used to divert room air around the secondary coil. A simple filter is included at the intake to the coil.

There are a number of variations of the induction unit system and these include:

- Two-pipe nonchangeover;
- Two-pipe changeover;
- Four-pipe nonchangeover.

To illustrate the operating principles, a two-pipe nonchangeover is described below.

Two-Pipe, Nonchangeover Induction System

A typical arrangement of a two-pipe, nonchangeover induction system is shown in Fig. 7.2. The secondary chilled water is supplied to the unit coils all the year round. In summer, the cooled and dehumidified primary air gives a measure of control of room humidity; these systems are unable to deal with

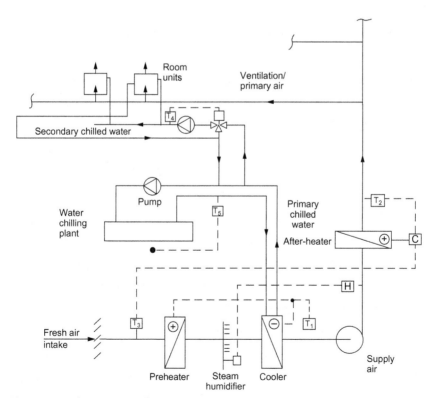

Fig. 7.2 Nonchangeover induction system.

large latent heat gains. In winter, the primary air is heated and humidified; this primary air supply takes care of the maximum heating requirements, but since this supplies only 15%–20% of the total flow rate from the unit, the maximum supply air temperature will need to be considerably higher than for constant flow, all air systems.

For efficient operation, a temperature schedule (based on the analysis of the building heating and cooling loads; see Chapter 18) is determined for the primary air supply, as shown in Fig. 7.3. There is some inefficiency since the primary air is heated to deal with the maximum heat loss and the secondary coil cools the induced air to meet the miscellaneous room sensible heat gains.

To prevent condensation on the secondary coils, the chilled water must be above the room air dew-point temperature (the drain pan on the coil caters for condensation of water vapour on start-up). To ensure this, the secondary coil water is taken from the piping circuit after the primary cooling coil in the fresh air plant.

Control

The primary air moisture content is determined by the air condition leaving the cooling coil; in summer, this is regulated through thermostat T_1.

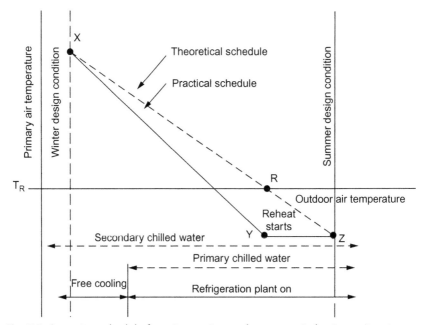

Fig. 7.3 Operating schedule for primary air, nonchangeover induction unit system.

The primary air is maintained at its scheduled temperature through the outdoor air thermostat T_3. The output of each room unit is usually regulated through a room thermostat, acting on a two-port valve.

Psychrometric Cycles
Summer Operation

A psychrometric cycle for summer operation is shown in Fig. 7.4. Outdoor air **O** is cooled and dehumidified to give the coil process **OB**. A temperature rise from fan and duct heat gains gives primary air temperature **P**. At the induction unit, the primary air induces the room air **R** through the secondary coil to cool the air to **R'**. This cooled secondary air then mixes with the primary air **P** to produce the supply air condition **S**, with a room ratio line (RRL), **SR**.

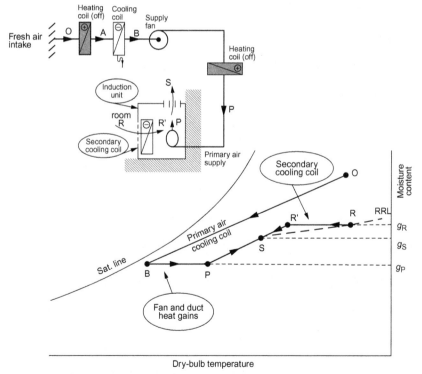

Fig. 7.4 Psychrometric process for nonchangeover induction system, summer operation.

Example 7.1

An induction unit has an induction ratio of 4:1. For the design conditions given below, determine:

(a) the primary air moisture content,
(b) the secondary coil load,
(c) the temperature of the air off the secondary coil.

Design Data

Room heat gains	
Sensible	7.5 kW
Latent	0.8 kW
Room condition	21°C, 60% sat
Minimum supply air temperature	14°C
Primary air cooling coil contact factor	1.0
Temperature rise in primary air due to fan and duct heat gains	3 K
Specific heat of humid air	$1.02 \, kJ/kg_{da} \, K$

Solution

Table values of moisture contents are used in the calculations. The results are shown in Fig. 7.5.

The supply air mass flow is obtained from Eq. (5.1):

$$\dot{m}_a = \frac{q_s}{c_{pas} \Delta t_c}$$

$$\dot{m}_a = \frac{7.5}{1.02 \, (21 - 14)} = 1.05 \, kg/s$$

Consider latent heat gain, using Eq. (2.21):

$$q_l = \dot{m}_a \, h_{fg} \, (g_R - g_S)$$

From tables, room moisture content $= 0.00943 \, kg/kg_{da}$

$$0.8 = 1.05 \times 2450 \, (0.00943 - g_S)$$
$$g_S = 0.00912 \, kg/kg_{da}$$

$$\text{Induction ratio} = I_r = \frac{4}{1} = \frac{g_S - g_P}{g_R - g_S}$$

$$\therefore g_P = 0.00788 \, kg/kg_{da}$$

For a primary air cooling coil contact factor of unity, the temperature of the air:

$$\text{leaving the coil} = 10.4°C$$

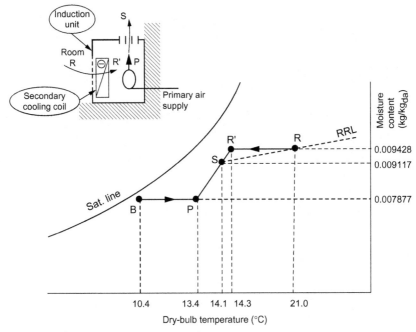

Fig. 7.5 Psychrometric process—Example 7.1.

Primary air supply to the unit $= 10.4 + 3.0 = 13.4°C$

Primary air mass supply rate $= 0.2 \times 1.05 = 0.21$ kg/s

Cooling provided by the primary air at the induction unit is given by:

$$Q_p = 0.21 \times 1.02\,(21 - 13.4) = 1.63\,kW$$

$$\therefore \text{secondary coil cooling load} = 7.5 - 1.63 = 5.87\,kW$$

$$\text{Induced air mass supply rate} = 1.05 - 0.21 = 0.84\,kg/s$$

$\therefore$ Temperature of the air off the secondary is given by:

$$5.87 = 0.84 \times 1.02\,(21 - t'_R)$$

$$\therefore t'_R = 14.1°C$$

The primary air cooling coil design load is given by Eq. (6.1). The *total* design load for the secondary coils is based on maximum *instantaneous* cooling load of all units.

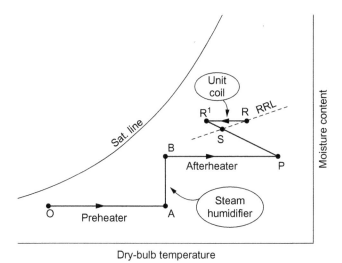

Fig. 7.6 Psychrometric process for nonchangeover induction system, winter operation (room with a net heat gain).

Winter Operation

A typical cycle for winter operation is shown in Fig. 7.6. The primary air is now heated and humidified to condition **B**. The reheater is operating to give the scheduled primary air temperature **P**. At the induction unit, the primary air induces the room air **R** through the secondary coil to cool the air to **R′**; **R′** mixes with the primary air **P** to give the supply air condition **S** and RRL **SR**. The secondary coil cooling load will depend on the level of miscellaneous heat gains offsetting the maximum heat loss; the line **PSR′** varies for each room.

FAN COIL UNIT SYSTEMS

In many ways, a fan coil system is similar to the induction system with changeover, nonchangeover, and four-pipe systems being used. The terminal units are usually mounted either under the window or in a ceiling void from which a number of supply outlets can be served; in the latter case, the ceiling void can be used for the return air.

The room units comprise of a fan, coil, supply and return air grilles, and filter (Fig. 7.7); with a four-pipe system, the coil is supplied with chilled and hot water. Room temperature is maintained through a thermostat either by controlling the fan speed or by a two-port valve on the room unit coil.

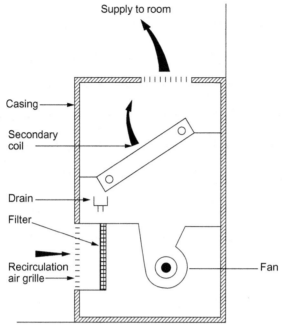

Fig. 7.7 Fan coil unit.

Where ventilation air is drawn through a grille in the external wall, condensation will occur on the unit coil, for which a drain tray and pipe is required; there may also be problems from dirt and noise penetration from the external environment.

A central air conditioning plant for fresh (outdoor) air supply is a more satisfactory solution for ventilation requirements. This allows greater flexibility in positioning the units and at the same time, as with the induction system, gives some control over room humidity. The fresh air can either be ducted to the terminal unit or supplied from an outlet, which is independent of the room unit. If the two supply air streams are discharged from separate points in the room, a psychrometric cycle can be drawn, as for the induction unit system in Figs 7.4 and 7.6.

UNITARY HEAT PUMP SYSTEM

Ronald W. James
Retired, previously principal lecturer at London South Bank University

The self-contained room units are reversible heat pumps designed for both cooling and heating. Each unit comprises a small refrigeration unit, a refrigerant/heat exchanger, a refrigerant/air coil, and a fan. The heat exchangers

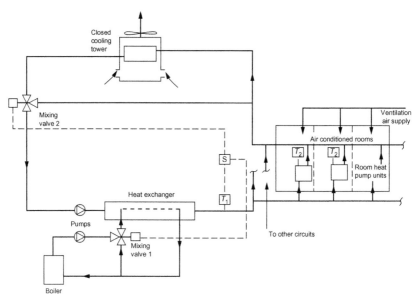

Fig. 7.8 Unitary pump system.

in each unit are connected to a water circuit, operating on a constant temperature, as shown in Fig. 7.8. This circuit includes a boiler and a cooling tower.

When the unit is required for cooling, the coil acts as the evaporator in the refrigeration circuit and the heat exchanger as the condenser.[1] For *heating,* the heat exchanger acts as the evaporator and the air coil as the condenser. This operation is achieved through a reversing valve (see Fig. 7.9) controlled by the room thermostat T_2. The refrigerant flow is reversed through all components except the compressor; the heat exchangers have to be designed for satisfactory operation in either mode.

This unitary heat-pump system is sometimes termed a heat reclaim system; however, this is not strictly true, since the piping circuits connecting the units do not allow the collection and storage of hot water off the room units. It should more correctly be termed a heat balance system, thermal balance being achieved when half the units are heating and the other half are cooling. In this state of operation, neither the boiler nor the cooling tower would be operating. When the system is not in balance, either the boiler would supply the out-of-balance load or the cooling tower would have to reject it.

---

[1] For a more detailed description of refrigeration systems, see Chapter 9.

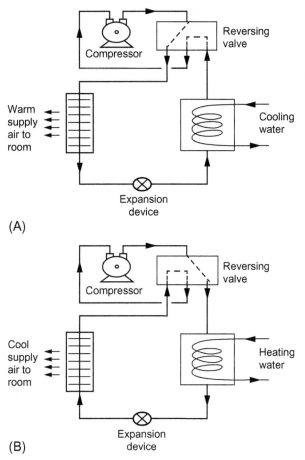

Fig. 7.9 Operation of reversing valve in room heat pump unit. (A) Heating of room air and (B) cooling of room air.

The system can serve perimeter and central core areas, and zoning is not required. Fresh air for ventilation can be supplied either through the external wall or from a central plant.

VARIABLE REFRIGERANT FLOW/VOLUME (VRF/VRV) SYSTEMS

Terry C. Welch

Retired, previously senior lecturer at London South Bank University

Historically, direct refrigerant cooling (where the space cooling coil is the evaporator of the refrigerant circuit) began as single circuits or split systems with the standard four components of evaporator, compressor, condenser, and expansion valve. The compressor and air-cooled condenser were located outdoors, and the refrigerant suction and liquid pipes ran indoors to the room/space being cooled and connected to the evaporator and expansion valve. For many buildings, this meant a proliferation of outdoor condensing units, which could be noisy and often dripped condensate. Two developments changed this type of air conditioning system; the first was the four-way reversing valve incorporated into the refrigerant circuit, which enabled the indoor unit to operate either in cooling or heating mode, making it a more attractive option. The second was the introduction of the *variable refrigerant flow* (VRF) system (also known as *variable refrigerant volume* (VRV) system), which allowed more than one indoor unit to be coupled to a single outdoor condensing unit. Two- and three-pipe systems are available; most manufacturers offer both systems that will feature variable speed compressor technology, R410a refrigerant and a large capacity range of up to 100 kW with a single outdoor unit. Systems have a modular design with a wide range of indoor units, both ducted and nonducted. The compressors are variable speed through an inverter drive.

Two-pipe VRF

A two-pipe VRF system is the simplest system providing either heating or cooling to the occupied space though it cannot offer simultaneous cooling or heating on the same system; by careful attention to building zones and occupancy profiles, the designer can offer the client a degree of flexibility.

Three-pipe VRF

Three-pipe VRF systems offer simultaneous heating and cooling on the same system and are termed by some manufactures *heat recovery systems*. In simple terms, this is achieved by removing heat from one area and conveying it to another. The circuitry is more complex than a two-pipe system and incorporates one more refrigerant line, the high-pressure, hot gas pipe. The system also features 'changeover' control units that divert the refrigerant gases either to or from the indoor unit depending on which mode of operation the system has been selected. This system is useful in a building that has a wide range of internal heat gains and/or building orientation.

SELF-CONTAINED AIR CONDITIONING UNITS

Ronald W. James

Self-contained air conditioning units contain an independent refrigeration circuit for cooling, a fan, a filter, and in some cases a heater. They are usually installed under a window, using outdoor air for condenser cooling. Ventilation air is also supplied through the external wall, though an independent air supply can also be used. A drain must be provided to deal with air-side condensation of water vapour. Control is through a room thermostat that cycles the fan on-off or adjusts the fan speed; unit sizes have a capacity range of between 1 and 10 kW.

DISCUSSION

Terry C. Welch

Benefits of Chilled Water Systems

Compared to direct refrigerant systems, water is a cheap, readily available, environmentally-friendly fluid. The waterside surface heat transfer coefficient is improved, making for more compact heat exchangers. Distribution piping, fittings, and controls are easily designed, selected, and installed. Control of air temperature and humidity in the space or process can be achieved accurately and with stability over a wide range of cooling loads. A whole range of fan coil units is available for use with chilled water and chilled beam/ceiling applications providing energy saving applications.

Water chillers/heat pumps contain the entire refrigerant charge within the unit, which can be placed either outside the main building or in a plant room. This considerably reduces the risk of refrigerant leakage and eliminates leakage within the air conditioned space.

Benefits of VRF Systems

As VRF technology develops, the benefits of these systems increase. One key area is their flexibility; because of their modular nature, they are relatively easy to zone, and for a multiple occupancy/use building, this characteristic is invaluable. Controls used by manufactures have the ability to provide accurate and specific control of temperature in each area of the building and thus provide efficient part-load control of the system.

One outdoor condensing unit can serve up to 32 indoor units (which could be a zone such as a floor or a wing of a building), and this can be controlled independently from the rest of the building. However, by rearranging the control, that zone can be incorporated with the rest of the building making one larger system. This larger system can be made up of multiples of the smaller ones, which can either be added or subtracted as the building usage evolves or expands in the future.

The installation of VRF systems is simplified because the manufacture will provide the consultant with a complete system that is based on the design criteria for the building. The early involvement of the manufacturer by the consultant ensures that the equipment selected has been correctly applied to the project and that a single point of contact is established from the outset. Equipment will be supplied and installed by one of the manufacturer's approved installers who are familiar with the equipment and the installation requirements. These installations take up less space than other systems, as the footprint of the outdoor units is smaller than for water chillers.

SYMBOLS

C_{pas}	specific heat of humid air
g	moisture content
h_{fg}	latent heat of evaporation
I_r	induction ratio
$\dot{m}_a$	mass flow rate of dry air
$\dot{m}_p$	mass flow rate of primary
$\dot{m}_r$	mass flow rate of recirculated air
Q_p	cooling load provided by primary air
q_s	sensible heat gain to air conditioned space
q_s'	sensible heat loss from air
q_l	latent heat gain to air conditioned space
q_l'	latent heat loss to conditioned space
t	dry-bulb temperature
$\dot{V}_p$	volume flow rate of primary air
$\dot{V}_r$	volume flow rate of recirculated air
Δt_c	temperature differential for cooling
Δt_g	moisture content difference

SUBSCRIPTS (FOR TEMPERATURE AND MOISTURE CONTENT)

P	primary air condition
R	room air condition
S	supply air condition

ABBREVIATIONS

RRL	room ratio line
VRF	variable refrigerant flow
VRV	variable refrigerant volume

CHAPTER 8

Chilled Beams and Radiant Ceiling Systems

Risto Kosonen

The average share of energy usage in the existing building stock is as high as 40% in the European Union (EU) member countries. To reduce the energy use in new buildings, the *Energy Performance of Buildings Directive* (EPBD) [1] requires that all new buildings in the EU should be *nearly zero-energy buildings* (nZEB) by 2020. Together with energy efficiency, there is a need to consider indoor climate conditions; chilled beam technology is one possible solution for future sustainable and nZEB buildings, and that is why chilled beams and radiant ceilings are popular systems for creating high-quality indoor climate conditions, including thermal comfort, with a low noise level, providing low energy consumption and competitive life-cycle costs.

In this chapter, the principles of chilled beam and radiant systems are described, along with typical applications and design principles.

COMMON APPLICATIONS AND BENEFITS OF CHILLED BEAMS

Chilled beam systems are used for cooling, heating, and ventilating spaces, where good indoor climate quality and individual space control are appreciated. These systems provide excellent thermal comfort, energy conservation, and efficient use of space due to the high heat capacity of water used as the heat transfer medium. Their operation is relatively straightforward and trouble-free, with minimum maintenance requirements. Chilled beams provide an option to integrate lighting and other building services with the room units, for example, lights, cabling, detectors, and controls.

There are two types of chilled beam system: *passive* and *active*. Passive beams comprise a heat exchanger for cooling, the operation based on natural convection with the ventilation air supplied to the space through separate diffusers. Also with passive beams, a separate space heating system is required.

Air Conditioning System Design
http://dx.doi.org/10.1016/B978-0-08-101123-2.00008-X

The most suitable applications for beam systems are those with the lowest air-side load fraction, because they are the ones that will benefit the most from the efficiency of hydronic systems. The operation of chilled beams is based on dry-cooling operation, and typically, the units do not include any piping for condensation. This means that the temperature of heat transfer surfaces must be higher than the dew-point temperature of the room air. For this reason, the humidity in these buildings has to be controlled, and in most cases, this means that the outdoor air has to be dehumidified in the central air-handling plant in warm, humid weather.

Typical applications for beam systems are the following:
- Cellular and open-plan offices
- Hotel rooms
- Hospital wards

Offices

In an office building, active and passive beam systems provide several benefits. The reduced supply airflow rate required from the air-handling system for ventilation purposes provides significant energy savings. The minimized supply air often translates to reduced reheat and fan power requirements. In addition, the smaller infrastructure required for this reduced airflow allows for smaller plenum spaces and plant rooms, translating into reduced floor-to-floor construction and increased usable floor space.

Hotel Rooms

Hotels can also benefit from active beam systems. Fan power savings often come from the elimination of fan coil units located in the occupied spaces. It also allows for the elimination of the electric service required for the installation of fan coil units and a reduction in the maintenance of the drain and filter systems. The omission fans from the occupied space also provide lower noise levels, which is a significant benefit in bedrooms.

Hospital Wards

For hospitals, the outdoor airflow rate required by local codes and guidelines is often greater than the requirements for cooling and heating loads, and also because of hygiene, it is important that the units are designed for easy vacuum cleaning; the water coil should be designed so that it can be removed for cleaning.

PASSIVE BEAMS

Heat transfer from passive beams occurs through natural convection; warm room air rises and comes into contact with the cold surface of the heat exchanger and then flows downward from the beam into the room; this is illustrated in Fig. 8.1. The beams can be positioned fully exposed, recessed within a suspended ceiling or above a perforated ceiling. Ventilation air from a central plant is not connected to the beams and can be introduced from either high- or low-level grilles.

Where the air is supplied using ceiling diffusers, the airstream should not obstruct the convective flow off the beam. If the beam is situated above the window or above other high convective heat loads, the possible reduction of cooling output should be taken into account. Passive beams should not be installed directly above fixed working positions, because the velocity created by natural convection is highest directly under the beam.

Passive beams can also be used with underfloor supply or with sidewall displacement terminals. This arrangement typically creates a satisfactory mixed flow system, where the convective downflow from the beam mixes with the low-velocity air supply.

Exposed Passive Beams

Exposed passive beams can help provide cooling in situations where there is a limited ceiling height whilst increasing the perceived room volume. When installing exposed passive beams, it is important to allow enough room above

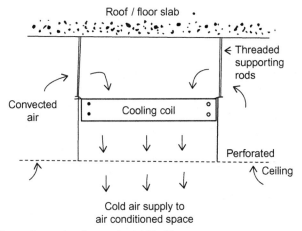

Fig. 8.1 Airflow schematic of a passive chilled beam.

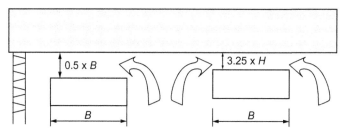

Fig. 8.2 The minimum distance of passive beam from the ceiling.

the beams for the recirculated air to flow freely. Passive beams typically provide specific cooling capacities between 150 and 250 W/m per beam length.
The capacity is influenced by the difference between the mean water temperature and the entering air temperature difference (Δt). The capacity
reduces as the difference between the average water and air temperatures
(typically $\Delta t = 8$–9 K) decreases that can also be influenced by the height
of the shroud,[1] beam configuration, and size. For example, doubling the
height of the shroud from 150 to 300 mm gives an increase of approximately
25% capacity at a Δt of 9 K.

If the space is insufficient, the cooling capacity will be reduced as the natural convective cycle is obstructed. The distance between the beam and the
ceiling should be at least 0.25 times the beamwidth (B) when there is an open
area on both sides. In the case where a passive beam is installed close to a wall,
the distance should be at least 0.5 B. The minimum distance of a passive
beam from the ceiling, close to a wall and open in both sides, is illustrated
in Fig. 8.2.

Recessed Passive Beam

Recessed passive beams may be installed behind a suspended ceiling as in
Fig. 8.1, though some local codes may not recommend this. In this application, a minimum clearance between the top of the beam and soffit should
be provided for a sufficient return air path. If possible, the return air path
should be placed adjacent to the unit; to maximize capacity, the net-free area
of return air path should be a minimum of 50% of the face area of passive
beam. If recessed beams are desired for aesthetic reasons and the interstitial
space cannot be used, it is possible to use a passive beam with an integral

---

[1] Shrouds are the side plates that create a 'chimney' and thus increase the cooling capacity of
the beam.

return path where the room air is able to pass through the coil with no contact with the ceiling void (plenum space).

Perimeter Passive Beams

Perimeter passive chilled beams are installed close to glazed façades or windows and are designed to offset solar gains in the perimeter zone. This beam arrangement minimizes the disruptive effect that solar gains can have on air temperature and air circulation away from the perimeter; this is especially important where displacement ventilation and/or chilled ceilings are used. A further advantage of locating chilled beams in the perimeter is that any warm plume rising from the window or blind enhances the air–water temperature difference in the chilled beam and this raises its cooling performance.

The thermal plume is trapped in the cavity formed between the beam ceiling and the slab above, directing the rising warm air through the coil. Passive beams mounted in this configuration may not directly affect the window temperature, but may offer increased performance due to the warmer air entering the coil. Care should be taken to fully explore this condition as the strength of the plume and soffit design may change the beam performance in unexpected ways. Perimeter passive beams can be integrated with an active chilled beam system or with an underfloor air distribution and radiant ceiling system.

Practical Guidelines to Passive Beam

Passive beams provide sensible cooling from the water system only. Heating and ventilation need to be handled by complementing systems. Selection of passive beams is crucial for capacity and comfort:
- Locating the passive beams in consideration of the natural movement of the room air can optimize beam output.
- The higher capacity per unit length, the more important the positioning becomes to maintain acceptable comfort.
- Use areas outside the occupied zone to allow the cooled air to mix with the room air before it reaches the occupants.

ACTIVE CHILLED BEAMS

Active beams are characterized by forced convection caused by the induction of room air across the water coil. This induction is created by primary air discharged through nozzles at high velocity (similar to the operation of induction system units described in Chapter 7), with a typical induction ratio

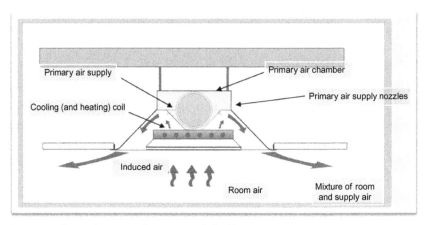

Fig. 8.3 Airflow schematic of an active chilled beam.

of about 1–4 and a nozzle pressure of about 50–150 Pa. Induced room air is then cooled (or heated) by the coil. The mixture of primary air and induced room air is supplied to the room through longitudinal slots along both sides of the beam. The airflows through an active beam are illustrated in Fig. 8.3.

Also available commercially are one-way discharge units and four-way discharge units with dimensions of 600×600 and 1200×600 mm.

Specific cooling capacity of active chilled beams is typically between 250 and 350 W/m of beam length. To ensure comfortable space conditions, it is recommended that the building is designed so that heat loads can be maintained below 80 W/m^2 of floor area; higher cooling capacities are possible, though there is an increased risk of draughts. Thus, the management of higher loads (80–100 W/m^2 of floor area) requires a more detailed analysis (e.g., mock-ups or CFD analysis) to reduce the risk of draughts.

Heating with Active Beams

When a chilled beam system is used for heating, satisfactory system operation cannot be achieved by oversizing the heating system. A new office building with a heating load of 30–45 W/m^2 of floor area is high enough to cover heat losses. If the heating inlet water temperature to a beam is higher than 40–45°C, the linear output of an active beam is higher than 140–160 W/m; the secondary air is often too warm to mix properly with the room air. The relatively low temperature gradient in the space raises the air temperature near the floor, thus maintaining comfortable thermal conditions and ensuring the energy efficiency of the system.

To prevent stratification from occurring, the heating water is supplied at temperatures that do not cause beam supply air temperatures of more than 9 K above the room temperature during occupied hours.

Practical Guidelines to Active Beam

Passive beams can provide both cooling and heating. Selection of passive beams is crucial for capacity and comfort:
- Cooling load is lower than 80 W/m^2 of floor area to guarantee satisfactory thermal conditions.
- Higher loads (80–100 W/m^2 of floor area) require more detailed analysis of thermal condition.
- In heating mode, the inlet water temperature should be lower than 45°C to guarantee less than 2 K vertical temperature gradient.

MULTI-SERVICE CHILLED BEAMS

The traditional chilled beam concept proposes an all-in-one solution for all ceiling-mounted room technical services such as ducts, lighting, sprinklers, sensors, power, and IT cabling. An integrated multiservice beam is suitable for both active and passive beams, both flush and exposed mounted installations. A service beam is an integration of aesthetics and economics. Pre-assembly of all services at the factory increases quality control whilst reducing costs. Single-source responsibility lowers risk and reduces the need for on-site coordination. In addition, the space achieves architectural finish with fewer separate pieces of equipment fixed to ceiling and walls.

DESIGN VALUES OF FOR CHILLED BEAMS

EN ISO 15251 (2007) [2] defines four categories of indoor environment requiring different ventilation rates based on the pollution rate generated by the used building materials. In offices, the required airflow rate is about 1–3 l/s per m^2 of floor area to maintain good indoor air quality level.

The chilled beam system primary airflow rate is determined to satisfy the comfort conditions, minimum ventilation requirement, and internal humidity level. In order to keep humidity levels within the design parameters, the primary air-handling unit normally requires the facility to dehumidify the supply air. The primary airflow rate should also be high enough to deal with the latent gains generated inside the space.

The ranges of environmental parameters for whole-body thermal comfort are prescribed in handbooks and standards (ASHRAE Handbook Fundamentals [3] ISO Standard 7730 2005 [4], and ASHRAE Standard 55 (2013) [5]). In practise, occupants may experience local thermal discomfort such as draughts, vertical temperature gradients, and radiant temperature asymmetry; cold/warm floors may also occur at the same time.

Together with the indoor climate targets, the heating and cooling capacity is the other major selection criteria of the chilled beam units. When the room unit is selected, other important considerations are as follows:

- Specific airflow rate (l/s per linear metre)
- Inlet water temperature
- Supply air temperature
- Water flow rate
- Air and water side pressure drop
- Noise level of the chilled beams

The recommended design values of chilled beam system are given in Tables 8.1 and 8.2.

Table 8.1 Recommended indoor climate values

Comfort parameters	Summer	Winter
PMV	$-0.5-+0.5$	$-0.5-+0.5$
Temperature		
Operative room air temperature	$24.5 \pm 1.5°C$	$22 \pm 2°C$
Vertical air temperature difference (0.1–1.1 m)	$<3°C$	$<3°C$
Radiant temperature asymmetry of windows	$<23°C$	$<10°C$
Radiant temperature asymmetry of ceiling	$<14°C$	$<5°C$
Floor surface temperature	19–26°C	19–26°C
Air quality		
Outdoor air requirement per floor area	$1.0–3 \, l/s,m^2$	$1.0–3 \, l/s,m^2$
Outdoor air requirement per person	8–20 l/s per person	8–20 l/s per person
Air velocity		
Draught rating (DR)	$<15\%$	$<15\%$
Average air velocity in the occupied zone (PMV = 0)	0.18 m/s	0.15 m/s
Maximum air velocity in the occupied zone	0.23 m/s	0.18 m/s
Air humidity		
Relative humidity	30%–55%	25%–40%
Acoustics		
Sound level requirement	NR15–NR30	NR15–NR30

Table 8.2 Recommended design values of chilled beams

Parameters	Cooling	Heating
Optimum heat loads/losses	60–80 W/m^2 floor area	25–35 W/m^2 floor area
Maximum heat loads/losses	<100 W/m^2 floor area	<50 W/m^2 floor area
Specific capacity of passive beam (above occupied zone)	<150 W/m	–
Specific capacity of passive beam (outside occupied zone)	<250 W/m	–
Specific capacity of active beams (highest class of indoor climate)	<250 W/m	<150 W/m beam length
Specific capacity of active beams (medium class of indoor climate)	<350 W/m	<150 W/m beam length
Supply air		
Specific primary airflow rate of active beam	5–15 l/s,m beam length	5–15 l/s,m beam length
Supply air temperature	16–20°C	19–21°C
Pressure drop of active beam	50–120 Pa	50–120 Pa
Room air		
Reference air temperature (air into the beam): Active beam	Room air temperature	Room air temperature 0–2°C
Reference air temperature (air into the beam): Passive beam	Room air temp. 0–2°C	–
Inlet water		
Water flow rate with pipe size of 15 mm (turbulent flow)	0.03–0.10 kg/s	0.03–0.10 kg/s
Water flow rate with pipe size of 10 mm (turbulent flow)	0.015–0.04 kg/s	0.015–0.04 kg/s
Inlet water temperature	14–20°C	30–45°C
Pressure drop	0.5–15 kPa	0.5–15 kPa

LOCATION OF CHILLED BEAMS

The recommended optimal location of active beams is above workplaces because the air velocity is lowest directly underneath the beam (if the throw pattern of the beam is horizontal). If the beam is positioned near a wall, the control of throw pattern or induction ratios is recommended to reduce air velocities at the work station close to the wall. It should also be noted that cabinets and partitions can have a significant influence on air distribution. In some cases, cabinets turn the airflow towards the workplace, again creating draughts. Manufacturers' performance data, software, models, CFD, and mock-ups can be used to assist with detailed design.

Whether active beams can be arranged in parallel or perpendicular to the façade primarily depends on the application and may affect ceiling coordination. Typically, in a parallel installation, the beam length is between 1.2 and 2.4 m; shorter beams increase flexibility, but it also increases pipe and duct connections. In a perpendicular beam installation, the typical length is between 2.4 and 3.0 m, though in practise the total length could be as much as 10 m.

The lowest velocity conditions for all seasons can be created when chilled beams are installed perpendicular in the space. This normally means that longer beams can be used, resulting in lower cooling capacity requirement per linear metre of beam. A perpendicular installation to the façade is also beneficial in intermediate seasons, when the window surface is still cold, but due to internal loads, cooling is required in the space. In a parallel installation, the cool supply air is discharged towards the cold window surface, increasing the velocity underneath the window. In the summer, when the window surface is warm, the thermal plume from the window affects air distribution, and the jet can turn before it reaches the window.

The thermal plumes from internal heat gains and warm surfaces have a significant influence on air distribution. The interaction of jets and convection flows is a complex issue; in a single office, the highest air velocity is close to the corridor. The return air increases air movement at floor level, increasing the risk of draught at ankle level. In an open office, the highest air velocity can be moved towards workers, close to the corridor.

Layout Design

One of the first considerations of layout design is the architectural requirements such as aesthetics and flexibility. In addition, coordination with the rest of the reflected ceiling plan is required, for example, lights, sprinklers, and smoke detectors. The layout has considerable influence on the horizontal air discharge in the space and should thus be taken into account at the design stage, as it depends on the room or module dimensions, intended use, and flexibility required.

A chilled beam system can also be designed as a self-regulating system without a room controller. The water inlet and outlet temperatures (typically 19–22°C) are selected close to the minimum room air temperature. This ensures that with minimum heat loads, the room is not becoming too cold. Once the heat loads start to increase, the room air temperature rises. As the temperature difference between the ambient room air and inlet water increases, the cooling outlet of the beam also increases. The system operation is reliable, and the room temperature is kept within the desired range.

Chilled Beam Systems

A typical active beam system for cooling, including a primary air system, will be the similar to the induction unit system described in Chapter 7 and illustrated in Fig. 7.2. The secondary chilled water circuit for the chilled beams is connected to the primary chilled water system and is continuously circulated using a three-way mixing valve for flow water temperature control. The water inlet temperature of chilled beams (14–18°C) is generally significantly higher than the supply water temperature for the air-handling unit's cooling coil for dehumidifying the primary supply air (7–9°C). However, to minimize energy consumption, it is recommended that a separate chiller is used for the chilled beams and air-handling unit. Also in cold and temperate climates, it is possible to use outdoor and thus cover major part of the cooling energy with free cooling (see Chapter 6).

Where chilled beams are also used for heating, the system has two separate water circuits, a low-temperature (35–40°C) circuit for the chilled beams and high-temperature circuit for the air-handling unit's heating coil.

RADIANT CEILING SYSTEMS

Static cooling systems (chilled ceilings), over the past 40 years, have proved themselves capable of delivering high levels of occupancy comfort at reduced running costs. Frenger designed, supplied, and installed the 'World's Largest Radiant Chilled Ceiling' system in 1962, the 175,000 m², 27-storey high Shell Oil headquarters, situated on the River Thames in London. This building was also the first fully sealed air-conditioned building in Europe and was revolutionary at its time as this chilled ceiling used the River Thames water to cool the building. This was achieved by pumping in cold water from upstream to a secondary heat exchanger, which in turn cooled with radiant cooling, and then depositing the warmer return water from the secondary heat exchanger downstream.

This installation is still operating after nearly 50 years and is a testament to the integrity of the product and to Frenger's design capabilities.

Since that time, the cooling requirements for a typical office environment have increased considerably; higher occupancy densities and IT equipment have fueled this increase. It became apparent in the mid-1990s that the cooling capacity of a traditional chilled ceiling was not sufficient to meet the increased heat gains, and consequently higher-capacity passive chilled beams were introduced into perimeter zones to offset the solar load generated at the building façade.

Although passive chilled beams provided the extra cooling at lower cost than a radiant ceiling, the perimeter aesthetics suffered due to the fin coil batteries requiring large-size perforations and percentage open area to allow buoyancy airflow to circulate.

However, Frenger saw the opportunity to take all the benefits from a traditional chilled ceiling for radiant cooling and to develop a 'hybrid' product solution that also had the cooling performance of passive beams coupled with radiant cooling to yield the same aesthetics associated with the traditional radiant chilled ceilings and, in turn, low air velocities for compliance with ISO 7730 European Standard [4], given that Frenger's hybrid chilled beams have a 35%–40% of the total cooling by radiation.

Traditional Chilled Ceilings: First Generation

Approximately 70% of the total cooling is by radiant absorption and the remaining 30% by convection if the backs of the tiles are insulated and about 55% radiation and 45% convection if the cooling tiles are uninsulated. The cooling tiles are constructed from zinc-coated steel that is polyester powder coated to whatever the project colour requirements are. Aluminium extruded heat exchange 'pipe seats' are powder coated black and are bonded to the back of the perforated metal tile. The tiles can be any size and as large as 1.35×1.35 m; these are known as 'Mega Tiles.'

There is a limiting factor of approximately 80 W/m^2 of activated ceiling tiles if insulated. Up to 90 W/m^2 of activated ceiling tiles is possible if ceiling tiles are uninsulated. The above listed cooling effects are based on 8.5°C difference between 'mean water temperature' (MWT) and the 'design room temperature', known as dt(K).

It should be noted that not all of the ceiling can be activated with cooling panels. In some instances, some ceiling grid can represent around 20% of the overall ceiling if the grid was 1.5 × 1.5 m and each plain grid was 150 mm wide. An allowance of approximately 8% of the total ceiling area being taken up by light fittings should also be made, and as such, the rule of thumb is that about 70% of the total ceiling area is to be activated by cooling panels. As such, 80 W/m^2 results in 57.6 W/m^2 on the floor (insulated tiles), and 90 W/m^2 results in 64.8 W/m^2 on the floor (uninsulated tiles) at 8.5 K difference between the mean surface temperature of the radiant panel and the room air temperature.

Common Applications and Benefits

Recently, radiant heating and cooling (RHC) systems have been widely applied, not only in residential but also in nonresidential buildings such as office buildings, retail stores, and schools and even large-scale buildings such as airport terminals and railway stations. Radiant floor heating systems are installed in all climate conditions; furthermore, with the combination of a ventilation system to handle latent load, radiant cooling systems have been shown to be feasible in hot and humid climates.

As with chilled beam systems, the radiant cooling system should be designed to prevent condensation on radiant surfaces. The inlet water temperature must be selected to avoid condensation, and the supply airflow rate should be high enough to remove internal humidity loads.

For technical guidelines, the ASHRAE Handbook [6] deals with the principles of radiant systems, heat transfer calculation, general design consideration, electrically heated radiant systems, design procedure, and controls. The REHVA Guidebook [7] includes system types, heating/cooling capacity, control, operations, and installation of low-temperature heating and high-temperature cooling systems. Recent guidelines for the TABS provide information on the design, commissioning, operation, and control of GEOTABS, which is a geothermal heat pump combined with a thermally activated building system [8].

Concepts of Radiant Heating and Cooling

The RHC system is defined as a system in which radiant heat transfer covers more than 50% of heat exchange in a conditioned space. Compared with all-air systems that depend on convection only, the RHC system provides heating and cooling by a combination of radiation and convection.

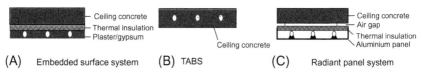

Fig. 8.4 Classification of the radiant heating and cooling system.

Depending on the position of the piping in the building, an RHC system is classified as follows:

(A) Embedded surface systems (pipes placed within a building layer (floor, wall, and ceiling)) that are isolated from the main building structure

(B) Thermally activated building system where pipes integrated into the main building structure (ceiling, wall, and floor) (TABS)

(C) Radiant panel system (pipes integrated into lightweight panels)

A classification of the radiant heating and cooling system is shown in Fig. 8.4 [9].

THE INDOOR CLIMATE OF RADIANT HEATING AND COOLING SYSTEMS

By utilizing the radiant heat transfer between the human body and radiant surfaces, a radiant heating system can achieve the same level of thermal comfort at a lower air temperature than a conventional system. A radiant cooling system can provide the equivalent thermal comfort at a higher air temperature and thus has energy-saving potential. Due to the small temperature differences between a heated or cooled surface and the occupied space, the radiant heating and cooling system (RHC) can benefit from the self-regulating effect, which can provide a stable thermal environment for the occupants within the space.

A chilled ceiling combined with mechanical ventilation can significantly decrease the draught risks by reducing the vertical drop of an air jet, which is often observed in all-air systems. In addition, the RHC system can provide an ideal vertical air temperature gradient and mitigate cold draught due to excessive air movement, because it operates using significantly less airflow compared with conventional air systems. Even though the RHC system can minimize the risk of draught due to excessive air movement, it can cause a sensation of stagnant air, which makes occupants require more air movement. For this reason, it is necessary to ensure the appropriate air movement (or ventilation performance) whilst reducing the draught risk.

Although the RHC system can provide improved thermal comfort by utilizing radiant surfaces, the floor surface temperature needs to be carefully considered to prevent the local discomfort because the feet are in contact with the floor. Particularly for radiant floor cooling, provided the floor surface is not less than 19°C, the indoor thermal environment conforms well to comfort standards.

The room acoustics need to be considered when designing RHC systems; room acoustics can deteriorate with TABS systems since they are generally installed without any sound-absorbing material so that the thermal capacity may be maximized. In some cases, it is necessary to add free-hanging sound absorbers in the room space.

Heating and Cooling Loads

The capacity of the RHC system is determined by the heat transfer of each element in the system; this depends on the heat exchange between the radiant surface and the occupied space (convective and radiant heat exchange coefficients), the heat conduction between the surface and the tubes (surface material, type of concrete, type of piping system, slab thickness, and tube spacing), and the heat transport by water (water flow rate and temperature difference between supply and return). It is recommended that the manufacture's technical date is used when available.

The radiant heat transfer coefficient can be considered about 5.5 W/m^2 K, whilst the convective heat transfer coefficient can vary between 0.3 and 6.5 W/m^2 K, depending on the position and temperature of the heated or cooled surface. In practical applications, the heating and cooling capacities of an RHC system are determined based on the following equations from REHVA Guidebook [8]:

Floor heating and ceiling cooling:

$$q = 8.92 \left(t_o - t_{sm}\right)^{1.1} \tag{8.1}$$

Wall heating and wall cooling:

$$q = 8\left(t_o - t_{sm}\right) \tag{8.2}$$

Ceiling heating:

$$q = 6\left(t_o - t_{sm}\right) \tag{8.3}$$

Floor cooling:

$$q = 7\left(t_o - t_{sm}\right) \tag{8.4}$$

where
t_o is the operative temperature of the space and
t_{sm} is the average surface temperature of the radiant panel.

Control of RHC Systems

Both supply water temperature and water flow rate are commonly used for the RHC system controls. In particular, attention should be paid to control the radiant cooling system to prevent surface condensation. To do this, the supply water temperature should be controlled so that the surface temperature does not become less than the highest dew-point temperature in the conditioned zones.

Due to the thermal mass of the radiant structure, the continuous operation with a water temperature that is too low or too high can result in undercooling or overheating problems. Radiant systems with large surface areas and small temperature differences between surfaces and the occupied space have a significant level of self-regulation. Small changes in room temperature will significantly change the heat exchange.

ABBREVIATIONS

EBPD	energy building performance directive
CFD	computational fluid dynamics
GEOTABS	the combination of thermally activated building system (TABS) with a ground heat exchanger and heat pump
nZEB	nearly zero-energy building
MWT	mean water temperature
PMV	predicted mean vote
RHC	radiant heating and cooling system
TABS	radiant heating and cooling systems with pipes embedded in the building structure (slabs and walls)

REFERENCES

[1] Energy Performance of Buildings Directive, EPBD Directive 2010/31/EU of the European Parliament.
[2] EN ISO 15251, Criteria for the Indoor Environment Including Thermal, Indoor Air Quality, Light and Noise, European Committee for Standardization, Brussels, 2007.
[3] ASHRAE Handbook Fundamentals; American Society of Heating, Refrigerating and Air-Conditioning Engineers, Inc., Atlanta.
[4] ISO 7730, Ergonomics of the Thermal Environment-Analytical Determination and Interpretation of Thermal Comfort Using Calculation of the PMV and PPD Indices and Local Thermal Comfort Criteria, International Organization for Standardization, Geneva, Switzerland, 2005.
[5] ASHRAE Standard 55 2013, Thermal environmental conditions for human occupancy. American Society of Heating, Refrigerating and Air Conditioning Engineers.
[6] ASHRAE Handbook: HVAC: Systems and Equipment, American Society of Heating, Refrigerating, and Air Conditioning Engineers, 2016.
[7] J. Babiak, B.W. Olesen, D. Petras, Low Temperature Heating and High Temperature Cooling: REHVA Guidebook No. 7, REHVA, Brussels, 2009.
[8] F. Bockelman, S. Plesser, H. Soldaty, Advanced System Design and Operation of GEO-TABS Buildings: REHVA Guidebook No. 20, REHVA, Brussels, 2013.
[9] J. Feng, S. Schiavon, F. Bauman, Cooling load differences between radiant and air systems, Energy Build. 65 (2013) 310–321.

CHAPTER 9

Refrigeration and Heat-Pump Systems

Ronald W. James*, Terry C. Welch†
*Retired, previously principal lecturer at London South Bank University
†Retired, previously senior lecturer at London South Bank University

Air conditioning usually implies that mechanical cooling is required, necessitating the provision of refrigeration systems. The cooling coils in the air conditioning plant may be supplied with either refrigerant or chilled water. Whichever is appropriate, the design of the refrigeration system requires detailed consideration if it is to match the air conditioning demands for cooling and achieve energy-efficient operation. At the same time, some of the heating requirements can be met using heat rejected by the refrigeration systems, or the system may be used purely for heating, as in a heat pump.

With power cycles such as those involving boilers, steam turbines, and condensers, heat is received by the working fluid at a high temperature and rejected at a low temperature, whilst a net amount of work is done by the fluid. The prime purpose of a refrigerator is to extract heat from a space, whereas a heat pump's principal purpose is to supply heat at an elevated temperature. The refrigerator and heat pump are identical in principle, and one machine is sometimes used to fulfil both functions.

PERFORMANCE CRITERIA

An adiabatic process is one in which there is no heat transfer between the fluid and its surroundings; a reversible process is an ideal process in which the fluid is in equilibrium at all times and is analogous to a frictionless process in machines. A process that is reversible and adiabatic is said to be isentropic.

A cycle is reversible if it consists only of reversible processes. The original concept of this was introduced by Sadi Carnot, who conceived a cycle in which all heat enters a system from a source at constant temperature, and all heat leaving it is rejected to a sink at constant temperature. It follows that the processes in which heat is exchanged with the surroundings must be isothermal, the temperature of the fluid never differing by more than an infinitesimal amount from the fixed temperature of the source or sink. When

Air Conditioning System Design
http://dx.doi.org/10.1016/B978-0-08-101123-2.00009-1

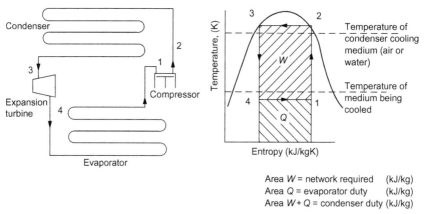

Area W = network required   (kJ/kg)
Area Q = evaporator duty   (kJ/kg)
Area $W + Q$ = condenser duty (kJ/kg)

Fig. 9.1 Reversed Carnot refrigeration or heat-pump cycle.

reversed, the cycle gives the highest possible efficiency for a refrigerator and is used as a basis for comparison.

A reversed Carnot cycle (Fig. 9.1) using a wet vapour as a working fluid has the following processes:

- Vapour is compressed isentropically from a low pressure and temperature to a higher pressure and temperature.
- The vapour is then passed through a condenser in which it is condensed at constant pressure.
- The fluid is expanded isentropically to its original pressure.
- The fluid is evaporated at constant pressure to its original state.

The criterion of performance of the cycle, expressed as the ratio output/input, depends upon what is regarded as the output. In a refrigerator, the *coefficient of performance*, COP, is defined as:

$$COP = \frac{\text{heat extracted in the evaporator}}{\text{net workdone on the fluid}} \qquad (9.1)$$

$$COP = \frac{\text{heat extracted in the condenser}}{\text{net workdone on the fluid}} \qquad (9.2)$$

From Eq. (9.2), it is evident that the COP of a heat pump is the reciprocal of the efficiency of a power cycle. A reversed Carnot cycle operating between an upper temperature T_a and a lower temperature T_b will therefore have a COP of $T_a/(T_a - T_b)$ for a heat pump and $T_b/(T_a - T_b)$ for a refrigerator. The Carnot COP cannot be achieved in practise but forms a convenient basis for comparison between cycles.

PRACTICAL REFRIGERATION CYCLES

It is possible to show that all reversible cycles operating between the same two reservoirs have the same COP, i.e., that of the reversed Carnot cycle. However, practical refrigeration systems operate on a different cycle that has a lower ideal COP but is more suitable in other respects. The expansion machine is replaced by a simple throttle valve referred to as the expansion valve; the expansion process is then one of the constant enthalpies instead of being isentropic. The compression process is carried out in the super-heated region. Finally, the condensed liquid is often cooled below its satu-ration temperature (subcooled) in the condenser or in the pipe connecting the condenser to the expansion valve.

The cycle is best demonstrated by using a pressure/enthalpy chart that is illustrated in Fig. 9.2. Since pressure is one of the coordinates of this diagram, the main family of curves comprises those of constant temperature. In the liquid region, these isothermals are nearly vertical because the effect of pres-sure on enthalpy is negligible; they run horizontally across to the saturated vapour line since, in the wet region of the chart, they are also lines of con-stant pressure. In the superheated region of the chart, they fall increasingly steeply. Also, on the chart, lines of constant volume are added in the wet and superheated regions, lines of constant dryness fraction in the wet region and lines of constant entropy in the superheated region. As with the psychromet-ric chart, a knowledge of two properties allows the others to be determined.

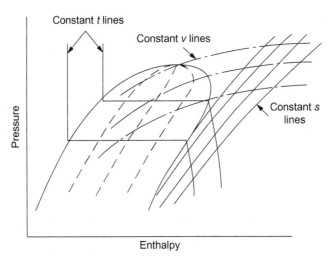

Fig. 9.2 Pressure-enthalpy diagram (*p-h* chart).

BASIC CALCULATIONS

The saturation cycle with isentropic compression is shown on a p–h chart, as in Fig. 9.3. Simple mass and energy balances give the following equations:
Refrigeration effect (kJ/kg)

$$Q_e = h_1 - h_4 \tag{9.3}$$

Compressor work done (kJ/kg)

$$W_D = h_2 - h_1 \tag{9.4}$$

Condenser duty (kJ/kg)

$$\begin{aligned} Q_c &= h_2 - h_3 \\ &= Q_e + W_D \end{aligned} \tag{9.5}$$

Refrigerant mass flow rate (kg/s)

$$\dot{m}_r = Q_e/(h_1 - h_4) \tag{9.6}$$

Compressor intake volume (m³/s)

$$\dot{V}_i = \dot{m}_r v_i \tag{9.7}$$

Compressor isentropic power (kW)

$$W_i = \dot{m}_r (h_2 - h_1) \tag{9.8}$$

Heat rejected in condenser (kW)

$$q_c = \dot{m} (h_2 - h_3) \tag{9.9}$$

Refrigeration duty (kW)

$$q_e = \dot{m} (h_1 - h_4) \tag{9.10}$$

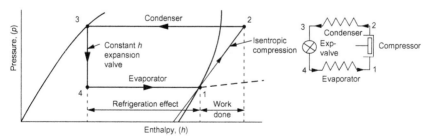

Fig. 9.3 Saturation cycle with isentropic compression.

Example 9.1

The system shown in Fig. 9.3 operates with refrigerant R134a with an evaporating pressure of 3 bar and a condensing pressure of 13 bar. The cooling coil load is 200 kW. Calculate the following:

(1) The COP when operating as a heat pump
(2) The COP when operating as a refrigerator
(3) The refrigerant mass flow rate
(4) Compressor intake volume flow rate
(5) Compressor isentropic power

Solution

The system shown in Fig. 9.3 assumes isentropic compression, wherein saturated liquid leaves the condenser and dry saturated vapour leaves the evaporator. From the pressure/enthalpy chart, Fig. 9.4

$$h_1 = 300.0\,\text{kJ/kg}$$
$$h_2 = 330.0\,\text{kJ/kg}$$
$$h_3 = 173.0\,\text{kJ/kg}$$
$$h_4 = h_3$$
$$v_i = 0.069\,\text{m}^3/\text{kg}$$

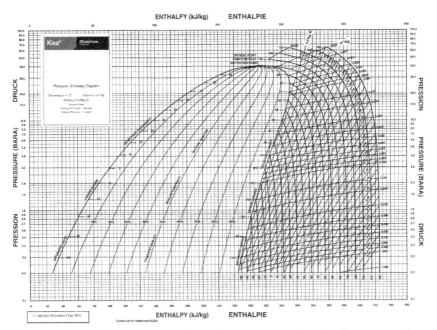

ENTHALPY (kJ/kg)   ENTHALPIE

Fig. 9.4 Pressure-enthalpy diagram for refrigerant R134a. *(Used with permission from Mexichem.)*

The enthalpy values will vary depending on the actual chart used, but this will not affect the results.

The refrigeration effect is given by Eq. (9.3):

$$Q_e = h_1 - h_3$$
$$Q_e = 300 - 173 = 127\,kJ/kg$$

The condenser duty is given by Eq. (9.5):

$$Q_c = h_2 - h_3$$
$$Q_c = 330 - 173 = 157\,kJ/kg$$

The work done by the compressor is given by Eq. (9.4):

$$W_D = h_2 - h_1$$
$$W_D = 330 - 300 = 30\,kJ/kg$$

(a) The COP for cooling is given by Eq. (9.1):

$$COP = \frac{\text{heat extracted in the evaporator}}{\text{net workdone on the fluid}}$$
$$\therefore COP = \frac{Q_e}{W_D} = \frac{127}{30} = 4.2$$

(b) The COP for heating (heat pump) is given by Eq. (9.2):

$$COP = \frac{\text{heat extracted in the condenser}}{\text{net workdone on the fluid}}$$
$$\therefore COP = \frac{Q_c}{W_D} = \frac{157}{30} = 5.2$$

(c) The refrigerant mass flow rate is given by Eq. (9.6):

$$\dot{m}_r = Q_e/(h_1 - s_4)$$
$$\therefore \dot{m}_r = 200\,kW/(300 - 173) = 1.57\,kg/s$$

(d) The compressor intake volume is given by Eq. (9.7):

$$\dot{V}_i = \dot{m}_r v_i$$
$$\therefore \dot{V}_i = 1.57 \times 0.069 = 0.108/s$$

(e) The compressor isentropic power is given by Eq. (9.8):

$$W_i = \dot{m}_r (h_2 - h_1)$$
$$\therefore W_i = 1.57(330 - 300)\,47.1\,kW$$

The effect of liquid subcooling in the condenser and vapour superheating in the evaporator is shown in Fig. 9.5. Liquid subcooling increases the

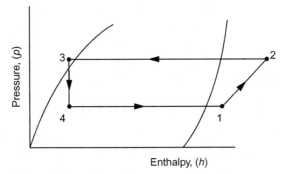

Fig. 9.5 Effects of liquid subcooling and vapour superheating.

refrigeration effect and therefore decreases the refrigerant mass flow rate, compressor-induced volume, and power for the same refrigeration duty. Vapour superheating in the evaporator increases the refrigeration effect and decreases the refrigerant mass flow rate for the same refrigeration duty, but there is a little change in the compressor-induced volume or power. However, when the vapour is heated in the tubes between the evaporator and compressor, the compressor-induced volume and power is increased.

Example 9.2

Repeat Example 9.1 assuming that the vapour entering the compressor is superheated by 10 K and the liquid entering the expansion valve is subcooled by 5 K, with all other conditions remaining the same.

Solution

From a pressure/enthalpy diagram:

$$h_1 = 310\,\text{kJ/kg}$$
$$h_2 = 344\,\text{kJ/kg}$$
$$H_3 = h_4 = 165\,\text{kJ/kg}$$
$$v_{i1} = 0.070\,\text{m}^3/\text{kg}$$

Using the appropriate equations:

refrigeration effect, from Eq. (9.3), is as follows: $Q_e = 145\,\text{kJ/kg}$
condensing effect, from Eq. (9.5), is as follows: $Q_c = 179\,\text{kJ/kg}$
compressor work done, from Eq. (9.4), is as follows: $W_D = 34\,\text{kJ/kg}$
COP (cooling), from Eq. (9.1), is as follows: COP $= 4.3$
COP (heating), from Eq. (9.2), is as follows: COP $= 5.3$
refrigerant mass flow rate, from Eq. (9.6), is as follows: $\dot{m}_r = 1.89\,\text{kg/s}$
compressor intake volume, from Eq. (9.7), is as follows: $V_i = 0.069\,\text{m}^3/\text{s}$
compressor isentropic power, from Eq. (9.8), is as follows: $W_i = 38.6\,\text{kW}$

Examples 9.1 and 9.2 show that, in theory, the influence of suction superheat and liquid subcooling on performance is small. In practise, small liquid droplets are entrained in the suction vapour, reducing the refrigeration effect; their influence decreases as suction vapour superheat increases.

SUCTION/LIQUID LINE HEAT EXCHANGERS

Suction/liquid line heat exchangers subcool the liquid refrigerant leaving the condenser by heating the vapour leaving the evaporator. Their influence on induced volume and power is usually small, but their inclusion in the system provides other advantages; if friction or change of fluid elevation of the evaporator over the condenser reduces the pressure of the liquid in the pipes between the condenser and expansion valve, it will boil and cause problems. Throttle valves are not designed to cope with vapour/liquid mixtures at entry. The heat exchanger prevents boiling by subcooling the liquid.

COMPRESSOR EFFICIENCIES

So far, isentropic compression has been assumed, but in reality, the compression will always absorb more power, so isentropic efficiency is the ratio of isentropic to actual work of compression. Leakage losses are incurred from high to low pressure. There will also be heat transfer to and from external sources to the compressed fluid.

Volumetric efficiency is the pumping ability of the compressor. It is the ratio of actual volume pumped to the displacement or swept volume of the compressor. It is mainly a function of the pressure ratio required for a particular application and is directly related to the evaporating and condensing temperatures at which the system operates.

SYSTEM DESIGN AND RELEVANT PLANT COMPONENTS

The diversity of systems makes it impractical to discuss fully their design or the plant components. The emphasis here is therefore one of presenting an overview appropriate to the air conditioning industry. The essential components of systems with, respectively, a flooded evaporator and a direct expansion (D-X) are shown in Figs 9.6 and 9.7.

The low-pressure region of the system in Fig. 9.7 is to the left of the dotted line and the high-pressure region to the right. If this system were a

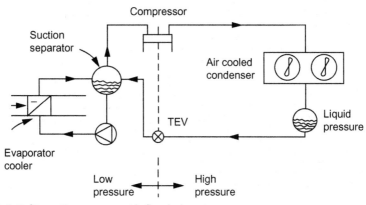

Fig. 9.6 Refrigeration system with flooded evaporator.

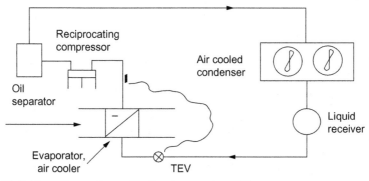

Fig. 9.7 Refrigeration system with direct expansion (DX).

boiler- and steam-turbine system, the essential components would be the same except that the expansion valve would be replaced by a liquid pump and the liquid receiver would be called a *hot well*. The significant difference is in the high- and low-pressure regions, which would be reversed. This provides a practical insight into the reason why refrigerators and heat pumps are the same machines and why, when reversed, the cycle becomes that of an engine.

Compressors

Reciprocating, screw, scroll, rotary vane, and centrifugal compressors are all used in refrigeration and air conditioning applications. Screw and centrifugal compressors tend to handle larger cooling capacities. Reciprocating compressors are not the dominant type used today, and the other types offer greater reliability and flexibility in the performance. Centrifugal compressors normally work with flooded evaporators and in air conditioning applications will form a package for water chilling.

The volume of vapour to be removed can be calculated as previously; the determination of the compressor swept volume requires the knowledge of its volumetric efficiency. For a reciprocating compressor, this is influenced by such things as clearance volume, valve size, valve inertia, valve spring tension, and effects of cylinder heating. The volumetric efficiency, η, can be expressed as a function of the ratio of condenser pressure to evaporator pressure, r. This may take the form:

$$\eta = A - Br^{\gamma} \qquad (9.11)$$

where the constants A, B, and index γ vary with compressor design and operating speed. A screw compressor's volumetric efficiency is a function of leakage between the rotors, heating through the rotors, and the pressure ratio.

Previously, it was shown that the COP improves if the difference between the condensing and evaporating temperatures is reduced. In addition, Eq. (9.11) shows that the lower the pressure ratio, the higher will be the compressor volumetric efficiency and therefore the smaller will be the compressor size. These are important fundamentals for the system design. As the condensing pressure rises, the compressor is required to do more work per kilowatt of refrigeration duty, and therefore, the capacity of a particular system is reduced. As the evaporator (and therefore compressor suction) pressure rises, less work is required for the same refrigeration effect. However, for a particular system, the increased suction pressure results in an increase in the density of the refrigerant and hence an increase in mass flow rate and capacity. The combined characteristic is shown in Fig. 9.8.

The capacity of a reciprocating compressor is controlled by manipulating the number of cylinders operating or varying its speed. Other methods are available such as suction gas throttling, but these are not energy-efficient.

Thermostatic Expansion Valve

Thermostatic expansion valves (TEVs) (or flow controllers) are designed to adjust the flow of refrigerant into the evaporator so that only vapour leaves the evaporator. At the same time, the liquid refrigerant proportion in the evaporator must be maximized to attain the highest heat-transfer efficiency

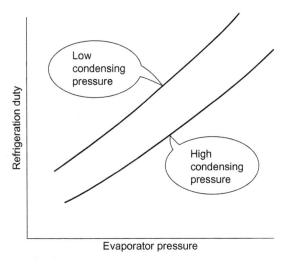

Fig. 9.8 Variation of refrigeration capacity with compressor suction and discharge pressures.

for the evaporator. This flow controller is widely used and responds indirectly to vapour superheat at the evaporator outlet. In its most usual form, the sensor consists of a small bulb or phial, containing the same refrigerant as the plant and in thermal contact with the evaporator outlet tube. It is connected to a flow regulator by a small bore tube. The regulator responds to the difference between the evaporator and phial pressures, which is related to the vapour superheat at the position where the phial is clamped to the pipe.

The majority of applications now use electronic flow controllers. They have two temperature sensors, one at the evaporator inlet for measuring the saturation temperature and one at the evaporator outlet for measuring the temperature of the superheated vapour. The refrigerant flow is controlled to maintain a constant difference between these two signals. They offer more accurate control of superheat and can also be incorporated into the control management system for the plant.

Packaged Water Chillers

Packaged water-chilling units, typically as shown in Fig. 9.9, are normally located on the roof of buildings. The size of these units ranges from 30 kW to an excess of 1000 kW. Chilled water pipes require insulating to prevent frost damage. Large rooftop units will have the refrigeration components outside the plant room, a disadvantage for servicing and

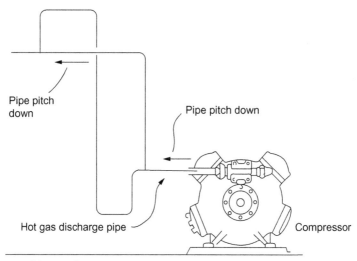

Fig. 9.9 Double riser from compressor to condenser.

maintenance. Their location within the plant room would require a large floor area and an additional cost for the supply and extract air ducts.

An alternative to the packaged water-chilling unit is to use a separate air-cooled condenser and liquid receiver. In this case, a packaged condensing unit would comprise a compressor, evaporator for cooling water, expansion valve, and associated piping and controls. Such a unit is usually located inside the plant room. These units require connecting pipes from the compressor discharge to the condenser and from the condenser to the expansion valve via the liquid receiver.

Package Condensing Units

These are available for direct coupling to cooling coils; they usually comprise a compressor, air- or water-cooled condenser, liquid receiver, and safety controls mounted on a base.

CONDENSING SYSTEM DESIGN

When the compressor capacity is reduced, the refrigerant velocity in the pipework is also reduced. This is unlikely to affect the expansion valve feed, but low velocities in the compressor discharge line can result in problems transferring oil carried over from the compressor to the condenser and from the condensation of refrigerant if the pipework is not designed correctly. A check valve (one-way valve) fitted in this line at entry to the condenser will prevent the condensation of refrigerant. Double risers are sometimes used as shown in Fig. 9.9.

When the capacity of the system is substantially reduced, the larger tube traps oil, and the refrigerant is then routed through the smaller tube. When the capacity is increased, the trapped oil is carried into the condenser allowing both pipes to function normally. A check valve is installed at the condenser entry to prevent vapour condensing in these pipes when the compressor is switched off.

The pipe connecting the condenser to the liquid receiver must be sized to ensure good draining. The flow should be part liquid and part vapour so that vapour can pass in either direction. The liquid velocity should not exceed 0.5 m/s.

Liquid refrigerant lines connecting the receiver and expansion valve must be free of vapour. A pressure reduction caused by friction or a change in static head will result in vapour being formed unless the liquid is subcooled. If the liquid receiver is mounted above the expansion valve and the pipes are correctly sized, the problem should not occur; if there is a problem, a suction/liquid line heat exchanger should be fitted.

Suction Lines

With any system in which the evaporators are remote from the compressor, the design of the piping between the evaporator and compressor is critical. A large pressure drop in this pipe reduces the system COP, whereas low refrigerant velocities allow the oil to be trapped. Generally, the suction vapour velocity should always exceed 6 m/s when the machine is operating at minimum capacity; if the compressor is located higher than the evaporator, the piping should be designed in a similar manner to that shown in Fig. 9.9.

Oil Separators

The use of a good oil separator reduces the problems encountered, but even if this has a high efficiency, some oil will still escape into the system; the design must cater for efficient oil circulation.

CAPACITY CONTROL

The capacity of a refrigeration system must be manipulated to match the load. Part-load efficiencies are particularly important for air conditioning applications in a temperate climate. The use of on/off control results in temperature fluctuations that are unacceptable for many applications, and on larger machines, frequent starting and stopping will cause damage to the system. Other capacity-control options available are as follows:

- Sliding vane for screw compressors;
- Variable speed compressor through inverter drive;
- Modulating inlet guide vanes for centrifugal compressors;
- Cylinder unloading;
- Back-pressure throttling;
- Hot gas bypass;
- Two-speed control.

Two-speed compressors provide energy-efficient capacity control but with large temperature fluctuations. Varying the compressor speed using an AC frequency inverter provides energy-efficient control and temperature accuracy, but the capital cost is higher. Cylinder unloading for reciprocating compressors is usually achieved using a mechanism that prevents the compressor suction valve from closing; this method is energy-efficient, but step changes in capacity result in temperature fluctuations.

Back-pressure throttling is achieved by restricting the refrigerant flow at compressor inlet. The energy consumption is reduced as the capacity is reduced, but the system is much less efficient than with the control systems previously discussed. The system is simple though the range through which the capacity can be varied is limited.

Hot gas bypass is a system by which some of the vapour discharged by the compressor is bypassed into the evaporator. The system provides the evaporator with an artificial load. Smooth capacity control from 0% to 100% load is possible, and high-temperature accuracy can be achieved. However, the energy consumption is almost unaffected by capacity changes, and the system is therefore inefficient.

COOLING TOWERS AND WATER-COOLED CONDENSERS

An alternative design of the condensing system is to use a cooling tower and a water-cooled condenser; the condenser is then much smaller and can be located in the plant room. The condensing pressure is generally much lower because the cooling tower performance depends on the ambient wet-bulb temperature. The net result is reduced energy costs but increased capital and maintenance costs.

Since public awareness that Legionnaires' disease is often traced to cooling towers, there has been an increase in the use of dry or sensible water coolers for condensers. This has resulted in bulkier condensing equipment, higher condensing temperatures, and reduced COP. A development for air conditioning applications with winter cooling requirements such as computer suites has been to use a glycol solution for condenser cooling, which can also be used directly in a cooling coil during winter. This saves compressor energy during such periods.

In water-cooled condensers, the refrigerant should condense at 5–10 K above the water inlet temperature, a low value results in a not only higher COP but also higher initial costs. With open-type cooling towers, the water should cool to within 3–6 K of the air wet-bulb temperature, and with closed cooling towers, the water should be cooled to within 8–12 K of the air wet-bulb temperature. (Cooling towers are described in Chapter 10.)

CHOICE OF COOLING SYSTEM

Design engineers often have to choose between D-X evaporators (Fig. 9.7) and a water-chilling system (Figs 9.10 and 9.11). Within D-X systems, there are *split systems* employing a single condensing unit and a D-X evaporator. There are also multiple D-X air-cooling evaporators connected to a single air-cooled condensing unit, known as variable refrigerant volume or

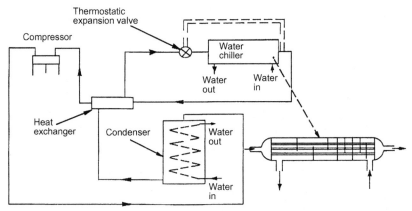

Fig. 9.10 Water chilling system with a dry expansion evaporator.

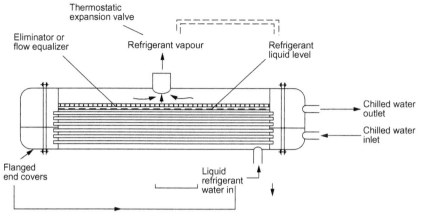

Fig. 9.11 Alternative flooded evaporator for water chilling.

variable refrigerant flow systems. All D-X systems are available as cooling/heating mode systems, where the cycle incorporates a *reversing valve*, enabling the evaporator and condenser to change their role. The language used now tends to speak of the outdoor and indoor units.

Chilled water systems will usually deal with large capacity, multistorey buildings. For many applications, the choice is difficult, and the final system may owe much to familiarity with particular equipment. Also, water chillers are often chosen to allow the design to be processed easily and quickly without the involvement of a refrigeration company.

The choice between a D-X system and the use of a heat-transfer medium should be made by considering the various factors particular to each application. The case for and against DX systems compared with water-chilling systems deserves consideration of the following:

For DX cooling coils,

- there is less pipework and equipment since there are no cold water circuits;
- higher coefficients of performance are possible;
- potential for reverse cycles or using the condenser to reheat air;
- avoiding problems associated with valve authority and valve characteristics in chilled water systems, in this way, total system capital cost is often reduced;
- energy-efficient capacity control over a wide range is possible;
- water chillers require protection against freezing during operation and during shut down;

- heat gains to chilled water in large-piped distribution systems; and
- lower off-coil air temperature can be achieved.

Against DX cooling coils,

- potential for refrigerant leakage into the occupied space;
- analogue control arguably more difficult than the use of a 3-way valve on a water/air heat exchanger;
- distribution of refrigerant piping to locations remote from the plant can cause problems, e.g., excessive pressure drop, leakage of refrigerant, and oil could be trapped;
- control of multiple coils through wide capacity ranges is more difficult;
- frost on coil surfaces is more likely to occur;
- water can be chilled without running the compressor for much of the year;
- increase in the quantity of refrigerant used;
- noise control is simplified by using a central water-chilling plant; and
- variable air volume flow rates (as in the VAV system described in Chapter 6) through the coil require special consideration.

The potential for refrigerant leakage from D-X evaporators has increased with the introduction of thin-wall copper tubing into their construction.

Water chillers with flooded evaporators (Fig. 9.11) can be designed to operate on a thermosyphon system to provide free cooling when the ambient wet-bulb temperature is low enough.

SYSTEM COMPARISON

It is informative to make a theoretical comparison between systems that can perform the same air conditioning process on the basis of refrigeration duty, required compressor swept volume, work done, and power requirements.

In this example, the air-cooling coil load is 100 kW. The following systems are considered to perform this air conditioning process.

- System 1

  An R134a DX evaporator cooling air, in which the refrigerant evaporates at 8°C and leaves superheated at 14°C.
- System 2

  An R134a water-chilling plant with a dry-expansion evaporator and an air/water heat exchanger.

For the dry-expansion water chiller, it is assumed that the water enters the chiller at 14°C and leaves at 7°C and that the refrigerant is evaporating at 2°C and leaves the chiller at 8°C. The refrigerant is assumed to condense at

Table 9.1 Comparison of refrigeration systems 1 and 2

	Chilled water	DX
Refrigerant mass flow	0.641 kg/s	0.625 kg/s
Power absorbed	23.1 kW	20.0 kW
Compressor displacement	0.049 m^3/s	0.037 m^3/s
COP	4.3	5.0

40°C and to leave the condenser at 35°C with both systems. To simplify the calculations, the heat transfer between all pipes and their surroundings, the influence of pressure drop in pipework and heat exchangers, and water-pumping power for system 2 is neglected. Assume an isentropic efficiency of 75% for compression, volumetric efficiency for the DX system of 92%, and for the chilled water 89%.

Calculations on the basic cycle, using Eqs (9.1)–(9.10), give the results in Table 9.1.

These calculations show that for indirect cooling there is an increase of ~24% in compressor size and 15% in power absorbed, when compared with the DX system.

Systems incorporating dry-expansion evaporators can benefit from suction/liquid line heat transfer. This applies to the system with the DX evaporator of Fig. 9.7 and to the dry-expansion liquid chiller of Fig. 9.10. Up to 40% of a dry-expansion evaporator may contain a mixture of liquid droplets and superheated vapour with a consequent poor heat-transfer coefficient.

If suction/liquid line heat transfer is used, a greater proportion of the evaporator will contain liquid in contact with the tube wall; this is achieved by sensing superheat between the heat exchanger and compressor to control the evaporator feed. This arrangement increases the risk of liquid refrigerant entering the compressor. By using computer control and an electronic expansion valve, superheat can be sensed at the heat exchanger outlet during normal operation and at evaporator outlet if the system transient behaviour results in risk to the compressor.

WATER CHILLING SYSTEMS

The water circuits for a water-chilling system with condenser heat recovery are shown in Fig. 9.12. The heating and cooling requirements are rarely matched. In winter, more heating duty is required, and in summer, more cooling duty is required. The system capacity is normally determined by

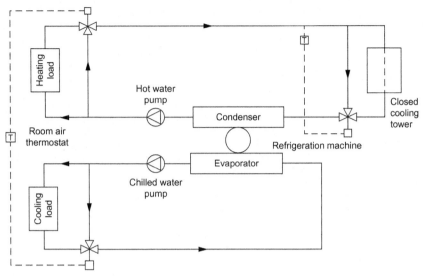

Fig. 9.12 Chilled water circuit with heat recovery.

the cooling requirements, and excess heat is dissipated in a cooling tower; additional heat is sometimes required, and this is usually provided by a boiler.

The system shown in Fig. 9.12 can be controlled to match the heating requirements, and for this, a second evaporator is installed in the ambient air to transfer additional heat when required. The cooling tower should be of the closed type so that any contaminants picked up are not circulated through the system.

Thermal storage can be introduced as shown in Fig. 9.13, to cope with peak loads, thus reducing the liquid chiller size.

An open cooling tower can be used in conjunction with a double-bundle condenser as shown in Fig. 9.14. Some of the condenser tubes are connected to the cooling tower and the remainder to the load.

When a water-chilling system is selected, design and installation costs can be reduced by purchasing packaged liquid chillers. This type of plant poses few problems to the system designer, but when more than one is employed, the running costs at part load may be higher than necessary. This results from some of the available evaporation and condensation surface areas not being utilized when a compressor stops.

An integrated system design could utilize all the heat exchanger surface area available at part load, reducing the temperature difference and increasing the refrigeration system COP. Nevertheless, separate systems matched.

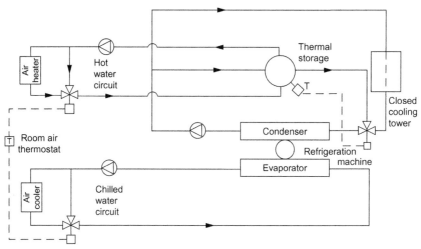

Fig. 9.13 Water circuits with thermal storage for condenser heat.

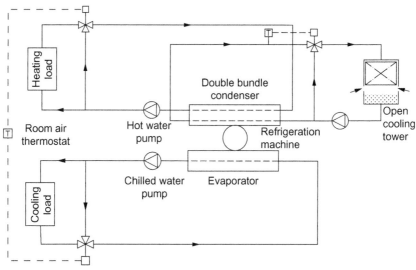

Fig. 9.14 Heat recovery system with double-bundle condenser.

In winter, more heating duty is required, and in summer, more cooling duty is required. The system capacity is normally determined by the cooling requirements, and excess heat is dissipated in a cooling tower; additional heat is sometimes required, and this is usually provided by a boiler.

The system shown in Fig. 9.12 can be controlled to match the heating requirements, and for this, an additional evaporator is installed in the ambient air to collect additional heat when required. The cooling tower should be of the closed type so that any contaminants picked up are not circulated through the system.

Thermal storage can be introduced as shown in Fig. 9.13, to cope with peak loads, thus reducing the liquid chiller size.

An open cooling tower can be used in conjunction with a double-bundle condenser as shown in Fig. 9.14. Some of the condenser tubes are connected to the cooling tower and the remainder to the load.

When a water-chilling system is the correct choice, design and installation costs can be reduced by purchasing packaged liquid chillers. This type of plant poses few problems to the system designer, but when more than one is employed, the running costs at part load may be higher than necessary. This results from some of the available evaporation and condensation surface areas not being utilized when a compressor stops.

*An integrated system design could utilize all the heat exchanger surface area available at part load, reducing the temperature difference and increasing the refrigeration system COP. Nevertheless, separate systems would be preferred with hermetic compressors, since a motor burnout would pollute all parts of the system including the remaining compressors.

REFRIGERANTS

The ideal refrigerant would be nontoxic and nonflammable and should have
* zero ozone depletion potential (ODP),
* zero global warming potential (GWP),
* short atmospheric lifetime.

These factors have become one of the main driving forces in selection of a refrigerant fluid, which should also have
* a large refrigeration effect requiring a small mass flow rate,
* a small amount of work to be done during compression,
* a small vapour specific volume.

These characteristics would result in a smaller compressor and a lower power requirement. Additionally, for reliable plant operation,
* the temperature of the discharge superheated vapour from the compressor should be low (generally 130°C is a maximum), to avoid oil breakdown;
* the evaporating pressure should be above atmospheric, and the condensing pressure should be below the critical pressure of the refrigerant.

Table 9.2 Refrigerants

Refrigerant	Type	ODP	GWP	Atmospheric lifetime (years)
R12	CFC	0.9	8500	102
R22	HCFC	0.06	1700	13.3
R134a	HFC	0	1300	14
R407C	HFC blend	0	1610	36
R410A	HFC blend	0	1900	36
Ammonia (R717)	Natural compound	0	0	<1
Propane (R290)	HC	0	3	<1
R1234yf	HFC unsat.	0	6	Very low
R1234ze	HFC unsat.	0	6	Very low

CFC, chlorofluorocarbon; *HCFC*, hydrochlorofluorocarbon; *HFC*, hydrofluorocarbon; *HFC*, blend consists of more than one HFC; *HC*, hydrocarbon; *ODP*, zero ozone depletion potential; *GWP*, zero global warming potential; *TEV*, thermostatic expansion valve.

All chlorofluorocarbon and hydrochlorofluorocarbon refrigerants are now banned for use in refrigeration equipment. The most environmentally friendly refrigerant fluids in terms of ODP and GWP are ammonia, propane, and blends of propane/butane. However, all these fluids have toxicity and flammability issues. Ammonia is used in air conditioning applications but requires a separate plant room and produces chilled water for cooling purposes. Propane is used in small-capacity air conditioning plants and has to meet current health and safety and flammability requirements.

There is currently research and development work with carbon dioxide as a refrigerant, and there are applications where it is used.

A group of fluids currently being developed are hydrofluoroolefins. They are propane-based fluids, unsaturated hydrofluorocarbons with refrigerant numbers HFC1234yf and HFC1234ze; they are being introduced to replace R134a. Propane is now used in centrifugal compressors and vehicle air conditioning. A summary of refrigerants is given in Table 9.2.

SYMBOLS

Q_e refrigeration effect
Q_c condenser duty
h enthalpy
$\dot{m}$ refrigerant mass flow rate
q_c heat rejected in condenser
q_e refrigerant duty
T absolute temperature
V_i compressor intake volume

v_i specific volume
W_D work done
W_i compressor isentropic power
x dryness fraction
η volumetric efficiency

SUBSCRIPTS

1, 2, 3, and 4   refer to specific conditions in the refrigeration cycle
a, b   temperatures

ABBREVIATIONS

COP   coefficient of performance
TEV   thermostatic expansion valve
VAV   variable air volume

CHAPTER 10

Humidifiers and Cooling Towers

The analysis of the humidity requirements of the occupants, building contents, building fabric, and manufacturing/industrial processes in the air conditioned space will determine the need for a humidifier in the system. The psychrometric processes associated with of humidifier have been given in Chapter 2. This chapter describes in more detail the humidifying equipment available, their operating characteristics, and guidance on practical application.

The second part of the chapter describes the various cooling towers available for the refrigeration systems explained in Chapter 9. For many years, mains water was used as a cooling medium for industrial applications and for condenser cooling requirements of refrigeration plants. However, advancing technology and man's increasing expectancy of his domestic, social, and working conditions have resulted in heightened demands for water, and this has meant that clean water is now relatively expensive. Consequently, where there is a cooling requirement, there are sound financial reasons to practise water conservation by using recirculatory cooling water systems. The most efficient plant available for this is the evaporative cooling tower.

HUMIDIFIERS

Spray Water Air Washer

The expression washer is a misnomer, the name originating from the days when this piece of equipment was used as an air-cleaning device in mechanical ventilation systems. In fact, it was found to be inefficient for air filtration requirements but was found to be an efficient adiabatic humidifier. Over many years, the water spray humidifier was developed into a sophisticated piece of air conditioning equipment in which various psychrometric processes described in Chapter 2. By using three banks of sprays, contact factors (defined in Chapter 2) of up to 0.96 can be obtained.

A typical water-spray-type washer is shown in Fig. 10.1. It consists of a chamber containing a set of nozzles that generate the fine spray of water, a

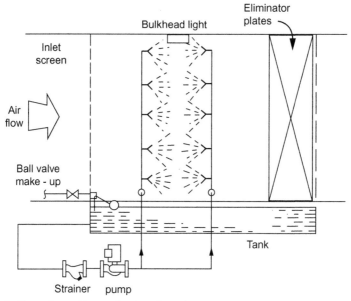

Fig. 10.1 Air washer with two banks of nozzles.

tank for the collection of unevaporated spray water, and an eliminator section at the discharge end that removes entrained droplets of water from the airstream, which would otherwise pass into the ductwork system. A pump recirculates water to the nozzles at a rate that is up to 50 times more than the evaporation rate.

The simplest design has a single bank of spray nozzles mounted in a casing usually 1–2 m long. Two or more spray banks are generally used when a high contact factor is required; the nozzles spaced to give uniform coverage across the airway. The casing, tank, and eliminator plates may be constructed from various alternative materials but most usually from sheet steel, galvanized after manufacture, for relatively small sizes; builders' work construction is usual for larger equipment. The eliminator plates consist of a series of vertical plates, 25 mm apart, with a number of bends and lips to promote the separation of water droplets from the air. An inlet screen may be provided to ensure even air flow through the washer. Other accessories include access and inspection doors, ball valve for water supply makeup, quick fill for rapid filling after maintenance, trapped overflow, drain, and bulkhead light.

Washers are commonly available from 1 to 100 m^3/s capacity, though there is no standardization of design parameters such as spray water density and spray pressure. Typical design values are

- face velocity,[1] 2.5 m/s;
- water circulation rate, 1.0 L/s per 1.0 m³/s air;
- pump pressure, 200 kPa;
- air pressure drop, 100 Pa.

Sprayed Cooling Coil

With the sprayed cooling coil humidifier, the water for humidification is sprayed through a bank of nozzles onto the air side of a chilled water coil or direct expansion coil. This arrangement avoids some of the problems associated with open-type air washers, e.g., scaling on the chilled water side of refrigeration evaporator tubes.

A diagram of a typical arrangement is shown in Fig. 10.2. The large wetted surface of the coil allows close contact between air and water, giving humidifying contact factors of up to 0.8. With no coolant being supplied to the cooling coil, the humidification process will be adiabatic. As explained in Chapter 2, when the coil comes into operation with a cooling load,

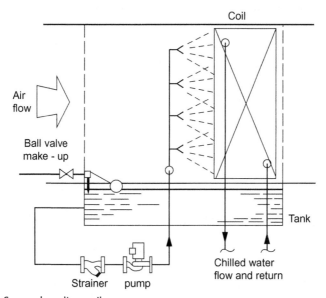

Fig. 10.2 Sprayed cooling coil.

---

[1] Face velocity is defined as the average velocity entering the equipment, i.e., the air volume flow rate divided by the cross-sectional area.

the off-coil condition will depend on the apparatus dew-point (ADP) temperature and the contact factor of the coil.

There are no standard design parameters; typical design values are the following:

- Face velocity, 2.5 m/s.
- Water circulation rate, 0.5 L/s per m^3/s air flow rate.
- Pump pressure, 100 kPa.
- Air pressure drop typically 200 Pa (depends on coil).

Capillary Washer

A diagram of a typical arrangement of a capillary washer is shown in Fig. 10.3. A bank of capillary cells is fitted across the airway, each cell comprising an open-ended box of galvanized sheet steel closely packed with glass fibres. Water sprayed onto one face of the bank of cells passes through to the other side, wetting the glass fibres. Air passing through the cells comes into close proximity with the large wetted surface area of the fibres, giving a high contact factor of up to 0.95. The water may be sprayed onto the upstream or downstream faces of the cells to give parallel or counterflow heat and mass transfer.

A capillary washer also acts as an efficient air filter, and as the cells accumulate dirt, the pressure drop increases and the contact factor deteriorates. It is therefore necessary to fit prefilters to prevent excessive accumulation of

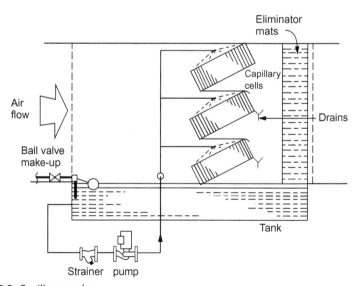

Fig. 10.3 Capillary washer.

deposits in the body of the cells. Even with filters, glass fibres tend to clog up, and an alternative to glass fibre is to use an open-mesh polyurethane foam packing.

Design values are similar to the sprayed cooling coil. Typical capillary cell dimensions are 600×600 mm across the face and 200 mm deep.

Steam Humidifiers

This group of humidifiers uses steam either from a central source, such as a boiler, or from a local generator close to the point of injection into the air duct. The advantages of these units compared with water spray types are the following:

- Smaller air pressure drop.
- Duct air velocity can remain high.
- Eliminator plates or mats are not required.
- Less risk of corrosion.
- No risk of bacterial contamination from standing water.

The steam injected into the air stream should be clean and dry. Drying the steam will minimize any odour that may be present in the steam supply from the boiler. A typical unit, as shown in Fig. 10.4, is designed to dry the steam

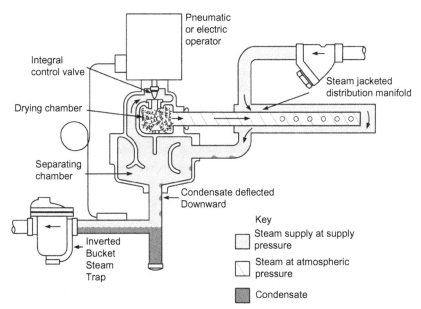

Fig. 10.4 Jacketed steam humidifier (for mains steam). *(Reproduced with permission from Armstrong International.)*

and to control its flow rate. The steam supply, reduced to the operating pressure, feeds into a separating chamber allowing any condensate to drain into a trap at the base of the unit. The steam flow rate is controlled by the valve at the top of the unit before passing into a drying chamber that is completely jacketed by the separating chamber. From here, the dried steam passes into a distribution manifold, which is itself jacketed by the incoming steam. The top of the drying chamber contains metal swarf to act as a sound absorber. The steam capacity of these units ranges from 0.15 to 72 g/s with a maximum operating pressure of 1.4 bar.

For local steam generation, water in a small cylinder is brought to the boil with electric resistance elements selected to ensure constant output irrespective of the conductivity of the water. Primary heating can also be provided by boiler steam or high-pressure hot water. The fresh steam produced passes over a series of baffles before being injected into the air stream via a stainless steel tray mounted in the duct. These units are available to provide a range of steam outputs, from 0.5 to 30 g/s.

Steam humidifiers are suitable for systems with air supply rates up to $10 \, m^2/s$. They take up less plant space than a water spray type and, if required, can more easily be incorporated into branches supplying different zones of the building. To avoid condensation on duct walls, the system should be designed and controlled to ensure that the air condition leaving the humidifier is less than 90% saturation. Where the humidifier is controlled from a room humidistat, a duct-mounted high limit humidistat should be placed in the duct downstream of the humidifier as a further precaution against condensation.

COOLING TOWERS

Stanley J. Marchant

Retired, previously senior lecturer at London South Bank University

A basic cooling tower is illustrated in Fig. 10.5. The principal components and their functions are as follows:

- A *case* encloses the cooling process and provides a structure for supporting the other items.
- The *packing* is a structure designed to provide the maximum area for the water to flow over; this maximizes the area of water air surface for heat and mass transfer. The materials for the packing used to be wood or metal slats, but these are now usually of plastic.
- A *fan* moves the required amount of air through the water-covered packing.

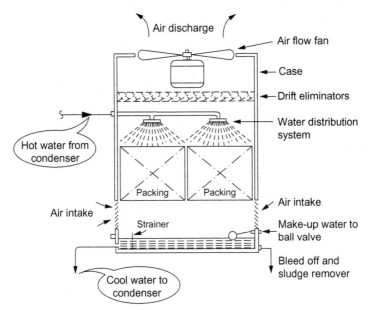

Fig. 10.5 Principle elements of a cooling tower.

- *Eliminators* are designed to minimize the carry-over and drift of water droplets from being carried away from the tower by the wind; this is particularly important in minimizing the risk of spreading the Legionella organism.
- A *water distribution system* ensures that the circulating water has maximum contact with the air. Nozzles produce fine water droplets and spread them uniformly over the packing.
- A *water reservoir* is an integral part of the casing, collecting the cooled water.
- A *pump* is required to provide the necessary pressure to circulate the water through the system; the water gravitates through the packing back into the reservoir.

Evaporation accounts for most of the heat transfer, and therefore, the process is referred to as *evaporative* cooling. In addition, depending upon the relative temperatures of the air and water, convection and conduction account for approximately a quarter of the heat transferred.

Open Cooling Towers

The type of tower shown in Fig. 10.6 is a mechanical, induced-draught, contraflow tower; it is named in this way because the fan draws a steady flow

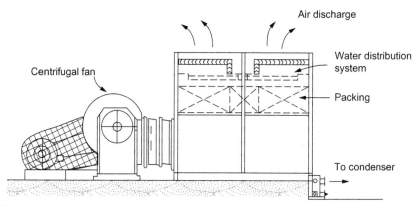

Fig. 10.6 Forced-draught contraflow with centrifugal fan.

of air through the packing in a flow direction opposite to that of the water, which falls by gravity. All the water used in the system is open, at some time, to the atmosphere; hence, the term *open* tower, to contrast with a *closed* tower, where the condenser water is in a closed piped water circuit.

Another type of contraflow tower is the *forced-draught* tower, two examples of which are shown in Figs 10.6 and 10.7. In both these cases, the air is discharged by the fan into the packing, but otherwise, the principles of operation are the same as for the induced pattern. Note the use of the centrifugal

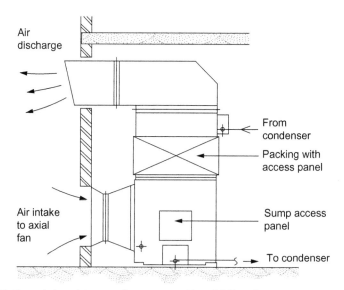

Fig. 10.7 Forced-draught contraflow tower with axial flow fan.

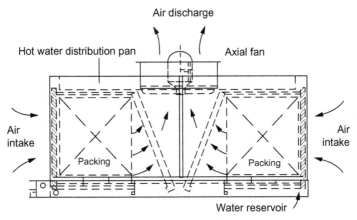

Fig. 10.8 Twin pack, induced-draught, cross flow tower.

fan in Fig. 10.6 and the axial flow fan in Fig. 10.7. Both types of fan are suitable for forced-draught application, but because of its configuration, the centrifugal fan is not normally suitable for an induced-draught tower.

Cross flow towers can be forced or induced-draught. The cooling principles are similar to contraflow types, but the air flows horizontally through the packing causing a cross flow through the falling water. The compact form of *induced* cross flow tower involves twin packs, one on either side of the fan as illustrated in Fig. 10.8. Because of the larger drift eliminators, there is less risk of carry-over with this configuration compared with a vertical tower, and low air velocities result in lower fan power and reduced noise levels. Heights are little more than that of the packing; consequently, a tower of this type is suitable for roof locations where there is an architectural need for a low profile.

An advantage of the cross flow tower is its simple water distribution system that consists of a shallow tank having a perforated case. Hot water from the refrigeration system condenser is directed from an open pipe into the tank situated immediately above the packing. This is in contrast to the contraflow towers where the air has to pass through the water distribution system consisting of spray nozzles on a trough and weir arrangement.

Closed Cooling Towers

Closed cooling towers are constructed on similar lines to open towers. The major difference is that the condenser water does not come into direct contact with the ambient air. Instead, a pipe coil is used through which the

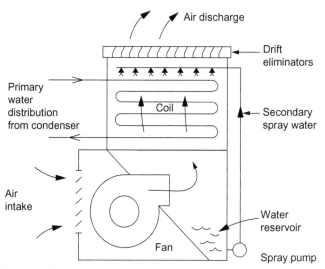

Fig. 10.9 Diagram of a closed cooling tower.

condenser cooling (primary) water is circulated. Secondary water, for evaporative cooling, is then sprayed onto the outside of the coil as shown in Fig. 10.9. For most applications, primary water can be cooled to within approximately 10 K of ambient wet-bulb temperature compared with approximately 5 K from an open cooling tower.

LEGIONAIRES' DISEASE

Legionella pneumophila occurs naturally in streams and stagnant pools and can be transmitted into building systems by way of the public water distribution network. Typical ranges of temperature of water in various building services, together with its effect on the bacteria, are shown in Fig. 10.10 [1].

The organism is harmless in low concentrations and at low temperatures but will multiply rapidly if ideal breeding conditions exist in the temperature range 25–45°C. As the temperature increases above about 47°C, the bacteria are killed at progressively increasing rates until at 70°C they die instantly.

Transmission of the bacteria requires aerosols to be generated from a contaminated water source. The most hazardous aerosol size is 5 μm; water droplets larger than this will settle out of the airstream, whilst smaller sizes are much less likely to carry the bacteria. Occasionally, inadequate maintenance and ineffective drift eliminators on some cooling tower installations have caused the dispersal of *Legionella* bacteria (in aerosol form) into the

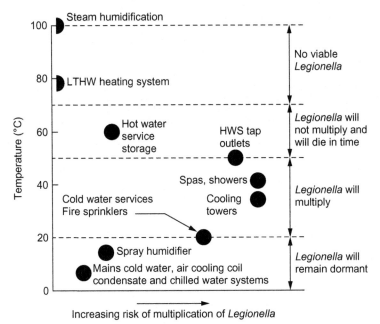

Fig. 10.10 Typical temperature ranges for water used in building engineering services, in relation to the *Legionella* bacteria. *(Reproduced from CIBSE Minimising the risk of Legionnaires' disease. TM13: 2013, with the permission from the Chartered Institute of Building Services Engineers.)*

atmosphere resulting in outbreaks of Legionnaires' disease. The adverse publicity surrounding these outbreaks has inevitably led to doubts about cooling tower technology and caused system engineers to reconsider the alternatives. One of these is the air-cooled condenser in which no water is used. This equipment is certainly hygienically safe, but other considerations make it a poor substitute for the evaporative cooling tower. By using only air *sensible* cooling results in an increase of approximately 30% of compressor power compared with a cooling tower.

With closed towers, the secondary spray water is exposed only to the atmosphere (a much smaller quantity of water than the primary water); it therefore presents a lower potential for dispersal of the bacteria than an open tower.

In the past, cooling towers have sometimes been underrated for the condensing requirements. The result of this is that the temperature of the water returning to the tower, from the condensers, will often be well above 25°C, a condition conducive to the multiplication of the *Legionella* bacteria. It is important that all spray-type towers should be sized so that the water in contact with the atmosphere will not exceed or rarely exceed this critical temperature.

There have been no reports of water spray humidifiers causing an outbreak of Legionnaires' disease. In a ducted air system, the aerosols should have dried out before reaching the conditioned space.

It is important to reiterate that eliminator plates on all water spray equipment should be designed to minimize carry-over of aerosols.

Water Treatment

With all recirculatory type spray equipment, bacterial growth (and possibly algae) will occur in the water tank or reservoir. Bacteria may also breed on cooling coil surfaces when these are wet due to dehumidification. As well as being a potential health hazard, this growth will reduce overall plant efficiency. Also, the buildup of salts in the recirculated water due to continued evaporation will affect heat and mass transfer and cause corrosion. Hence, water treatment is usually necessary to prevent the increase in bacteria and scale formation. Continuous bleed off or periodic discharge (blowdown), together with regular cleaning of the water tank and air side surfaces, is also vital.

The water treatment programme must be comprehensive and monitored by sample analysis, together with visual inspections as to general cleanliness.

In many areas, the normal supply water to a nonstorage adiabatic humidifier contains a relatively high proportion of carbonates, and when the aerosols generated by the humidifier evaporate, the solids remain in the airstream as a fine dust. This dust should be removed either by filtration after the humidifier or by water treatment which, to be effective, must remove the solids; simply to change their chemical composition is not sufficient.

REFERENCE

[1] CIBSE Minimising the risk of Legionnaires' disease. TM13: 2013.

CHAPTER 11

Exhaust Air Heat Recovery

A system that makes optimum use of recirculated air is the most efficient method of recovering heat from the exhaust air. However, if for ventilation purposes the system has to be designed for either 100% outdoor air or a large percentage of outdoor air, then consideration should be given to the incorporation of a heat recovery unit (HRU) for exhaust air heat recovery. Generally, the HRUs described below refer to the exhaust air-heating the supply air (the normal economic requirement in cold or temperate climates), but in most cases, the units can be arranged to provide some cooling during summer operation.

EFFICIENCY

The efficiency of an air-to-air heat recovery unit may be defined as follows:
Referring to Fig. 11.1:
For sensible heat exchange:

$$\eta_s = \frac{\dot{m}_s\left(t_S - t_O\right)}{\dot{m}_e\left(t_e - t_O\right)} \times 100 \tag{11.1}$$

For total heat exchange:

$$\eta_h = \frac{\dot{m}_s\left(h_S - h_O\right)}{\dot{m}_e\left(h_e - h_O\right)} \times 100 \tag{11.2}$$

where $\dot{m}_s$ is the air mass flow rate in the supply duct and $\dot{m}_e$ is the air mass flow rate in the extract duct.

THERMAL WHEEL

A thermal wheel comprises a framework, like a thick cartwheel, filled with a suitable matrix of large surface area through which the air can pass. The unit is installed between the two counterflowing airstreams as illustrated in Fig. 11.2. The wheel is rotated slowly, driven by a small electric motor with a maximum speed of 10 rpm. The part of the wheel in the exhaust air is

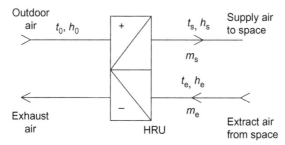

Fig. 11.1 Schematic of an air-to-air heat recovery unit.

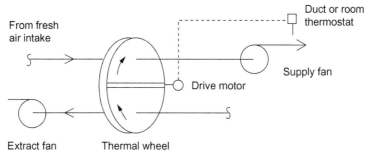

Fig. 11.2 Schematic of thermal wheel in the system.

warmed up, and this in turn heats the incoming air as the wheel revolves. The rate of heat transfer is regulated by varying the speed of rotation of the wheel (Figs 11.3 and 11.4).

Hygroscopic wheels transfer latent heat and sensible heat, and these will be particularly suitable for use in spaces that have high humidity. The psychrometric processes are shown in Fig. 11.5.

Exhaust air will not carry over to the supply air provided the correct pressure differentials are observed. To avoid leakage of the exhaust air into the supply airstream, the air duct pressure should be positive in the supply with respect to the exhaust, with the leakage path as indicated in Fig. 11.6. This is ensured by placing the supply fan on the upstream side of the wheel, but an excessive pressure differential across the unit will result in a flow of outdoor air to the exhaust with a reduction in overall efficiency. Where air flows from exhaust to supply, the efficiency is increased but with the risk of the exhaust air contaminating the supply air. A purge fitting, illustrated in Fig. 11.3, can be installed at the interface of the two airstreams to minimize this transfer.

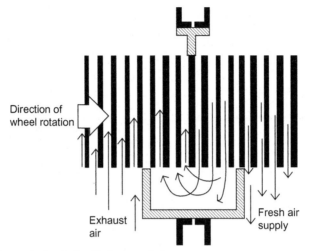

Fig. 11.3 Thermal wheel, detail of purge fitting.

Fig. 11.4 Thermal wheel. *(Photo courtesy of Fläkt Woods Limited).*

Typical design values for thermal wheels are:
- size, diameter of 0.5–4.5 m;
- air flow rate, range of 0.2–70 m³/s;
- pressure drops, 60–250 Pa;
- efficiency of heat transfer, 70–90%.

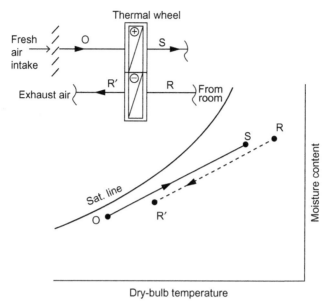

Fig. 11.5 Psychrometric process—hydroscopic thermal wheel.

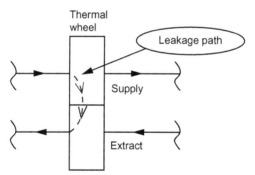

Fig. 11.6 Leakage path across a thermal wheel.

Some advantages of the thermal wheel are that efficiency remains high whatever the thermal load; there are no problems with bacterial growth and that frost/ice does not build up at subzero outdoor air temperatures on *sensible* heat exchangers (frosting can occur on hygroscopic wheels).

There may be difficulties in arranging the plant to obtain the required pressure differentials, and there is the need to bring exhaust duct close to the supply.

HEAT PIPES

The operating principle of a heat pipe is illustrated in Fig. 11.7A. A length of pipe, up to 3 m long and 50 mm diameter and sealed at both ends, contains a tightly fitting sleeve of porous material around a hollow core together with a charge of refrigerant. With one end in the warm airstream (W) and the other in the cold airstream (C), the refrigerant evaporates at (A) absorbing heat. The gas passes along the inside to condense at (C) giving out heat, and the refrigerant travels back by capillary action through the wick to complete the cycle.

In an air conditioning system, a bank of these heat pipes is used. For efficiency of heat transfer, the bank of pipes will be finned, the whole unit acting as a pair of conventional heat transfer coils. Efficiency is improved by tilting the tubes, and for conventional systems, this angle of tilt is about six degrees with the warmer end in the high-temperature airstream. Control of output from this unit may be achieved by face and bypass dampers in the supply air duct as shown in Fig. 11.7B. The heat pipe works in reverse for summer operation. The principal advantage of this unit is its relative simplicity resulting in low maintenance costs.

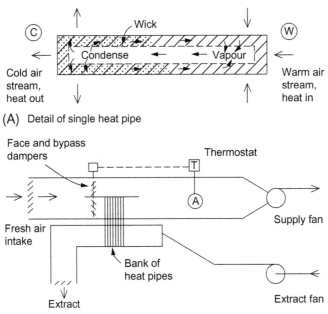

(A) Detail of single heat pipe

(B) Arrangement of heat pipe unit in system

Fig. 11.7 Detail of heat pipe unit and arrangement in a ducted air system.

Typical design values for heat pipes are the following:
- Air flow rate, range up to 5 m^3/s
- Face velocity, 2.5 m/s
- Efficiency of heat transfer:
  - 4-row coil, 50%
  - 8-row coil, 70%
- Pressure drop:
  - 4-row coil, 70 Pa
  - 8-row coil, 140 Pa

PARALLEL PLATE HEAT EXCHANGER

The principle of operation of the parallel plate heat exchanger is illustrated in Fig. 11.7. A basic unit comprises an open-ended box with a series of thin plates of metal, plastic, or glass. These form narrow linear passages, alternate rows of which carry the supply air, the remainder the exhaust air; the plates can be finned to enhance heat transfer. Some latent heat may also be recovered when the outdoor temperature is sufficiently low to condense moisture on the exhaust air side of the plates.

Units are obtainable in a large range of sizes and able to deal with air flow rates of up to 2 m^3/s; modules can be bolted together in parallel if higher system flow rates are required. Efficiencies are in the range of 60%–75%, and the output can be regulated with a modulating damper system as shown in Fig. 11.8, the thermostat maintaining the temperature at A (Fig. 11.9).

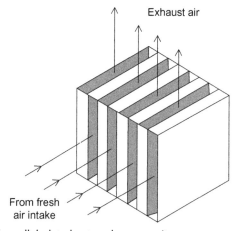

Fig. 11.8 Detail of parallel plate heat exchanger unit.

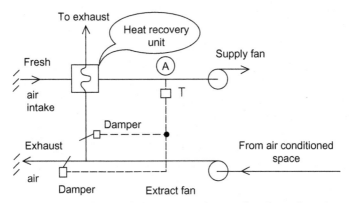

Fig. 11.9 Arrangement of parallel plate heat exchanger fitted in a ducted air system.

The advantages of this unit are that little maintenance is required and there is no possibility of cross contamination between the two airstreams. Apart from the need to bring the supply and extract ducts close together, disadvantages centre around condensation within the unit. In winter, this can cause the unit to ice up with consequent loss of efficiency and reduced flow rates. This type of HRU may therefore require frost protection, usually in the form of a preheater. If the unit is arranged for summer cooling, condensation can occur on the supply air side, and in these circumstances, there is a risk of bacterial contamination.

RUN-AROUND SYSTEM

Two pipe coil heat exchangers, one in the supply duct and one in the exhaust duct, are connected with a closed, pumped/piping system filled with water dosed with antifreeze. The exchanger in the exhaust air duct transfers heat to the water, which is circulated to the exchanger in the supply duct. A typical arrangement is shown in Fig. 11.10, the control of the supply duct temperature being achieved by a thermostat and three-way valve. The backup that may be necessary can be obtained by a heater in the piping system, rather than by a second air-heating coil.

The efficiency of heat transfer is in the range of 40%–70% depending on the design of the system. The pipework and fittings should be insulated to maintain maximum efficiency. In very cold weather, the liquid leaving the supply air coil can be cold enough to form ice on the exhaust coil, and when this is the case, a protective heating circuit will be necessary.

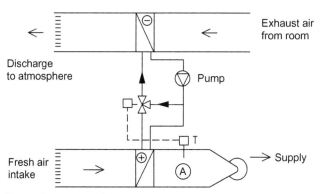

Fig. 11.10 Two-coil run-around system.

This method is used where the supply and exhaust ducts cannot easily be run close together; it is the most appropriate scheme for use where existing systems are to be upgraded. A number of coils may be incorporated, making this method a more flexible way of recovering heat from the exhaust air compared with other methods of energy conservation.

General

To save fan power, the unit can be bypassed when it is not providing heat to the system. For this to be effective, the flow rate handled by the fan should also be regulated.

In some systems, where the HRU is unable to meet the whole of the heating load, backup heaters will be required. If the air conditioning system is switched off, whether at night or at weekends, a conventionally sized heater will be required for start-up of the system. As with all heat exchange equipment, a filter should be placed upstream of the unit in both supply and exhaust airstreams. A preheater can be used to prevent condensation within the heat recovery unit where this could cause problems.

ECONOMIC ANALYSIS

When analysing the economics of an air-to-air heat recovery unit, the following cost differences should be considered, compared with a system designed without the unit:

Capital costs are as follows:
- Heat recovery drop;
- Associated controls;

- Ductwork connections;
- Filter in exhaust duct;
- Increase in fan size to deal with the additional pressure drops, supply and extract;
- Reduced heater and cooler battery sizes;
- Reduced boiler and refrigeration plant sizes.

Annual costs are as follows:

- Savings in energy for heating and cooling plant;
- Additional fan energy consumption;
- Energy consumption of ancillary equipment, e.g., motor for thermal wheel and pump for run-around system;
- Maintenance costs.

Most economic studies of units using 100% outdoor air show a payback period of between 1 and 5 years. The economics will be improved with higher exhaust air temperatures (when heating) and for systems with longer hours of operation. As well as comparing the different methods, it will usually be worth investigating two or three unit sizes since an HRU of the smallest size for a given air flow rate will not necessarily produce the shortest payback period.

An example of calculating the reduction of annual energy with an HRU is given in Chapter 18.

CHAPTER 12

Air Filters

Outdoor and recirculated air used in air conditioning and mechanical ventilation systems contain impurities and contaminants that need to be removed for the following reasons:

- To maintain an acceptable level of air purity for occupants, processes, and building equipment;
- To protect the air conditioning plant;
- To prevent staining on interior decorations and furniture.

In addition, airborne contaminant arising from any process occurring within the treated space may have to be removed from the exhaust air. Examples of this include grease in kitchen extract systems and wood dust in saw mills.

Atmospheric contaminants are classified as solid, liquid, gaseous, or organic, and usually, within these categories, they are ordered roughly according to particle size and type. The particle size is relatively small, and the unit of measurement commonly used is the micrometre ($\mu m = 10^{-4}$ m). Most of the staining in buildings is caused by particles with a size range of between 0.5 and 5 μm. Although invisible to the human eye, these particles have little mass but are very numerous. For comfort air conditioning applications, it is usually considered necessary to include a filter with efficiency high enough to remove most of the particles in that range.

FILTER PERFORMANCE

Efficiency

The most important characteristic of filter performance is its efficiency. Efficiency is a measure of the ability of the filter to remove dust from the air, expressed in terms of the dust concentration upstream and downstream of the filter; it is given by the equation:

$$\eta = \frac{C_1 - C_2}{C_1} \times 100 \qquad (12.1)$$

where $C_1 =$ upstream dust concentration and $C_2 =$ downstream dust concentration.

Air Conditioning System Design
http://dx.doi.org/10.1016/B978-0-08-101123-2.00012-1

Dust concentration can be expressed in terms of either mass, number of particles, or staining power per unit of surface area. Where the concentrations are expressed as a mass, the efficiency is known as the *arrestance* A.

There are a number of ways of determining filter efficiency using standard tests based on the way concentration is expressed, and as these give different values, it is important to state the test method used. For example, most filters whose efficiency is based on the mass of particles removed will have efficiencies of more than 90%, as the heavier, larger particles are more easily trapped by the filter medium; whereas based on a staining power test, the efficiency may be as low as, say, 20%.

There is no one test method suitable for all types of filters, and results from different methods cannot be directly compared. To compare the performance of various filters, the same test method must be used together with the same test dust or aerosol; e.g., the test dust should have the same particle-size distribution.

Tests of Filters for General Purposes

Tests for arrestance and dust spot efficiency are covered by BS 6540: part 1, [1] ASHRAE Standard 52–76 [2] and Eurovent 4/9—1997 [3].

The gravimetric tests use a standardized, synthetic dust with the specification given in Table 12.1. The test is carry out by feeding a measured, controlled quantity of test dust into the airstream, giving the value S of C_1 in Table 12.1. The mass of dust leaving the filter under test is determined

Table 12.1 Synthetic dust for testing filters (using standards ASHRAE 52-76 and Eurovent 4/9—1997)

Proportion by mass	Composition
72%	Natural earth dust, graded to give the following particle size by mass (tolerance ±3%) 0–5 μm 39% 5–10 μm 18% 10–20 μm 16% 10–40 μm 18% 10–80 μm 9% Typical bulk density—1057 kg/m^3
23%	Molocoo (carbon) black—a submicro Particle material forming fluffy aggregates Typical bulk density—272 kg/m^3
5%	No. 7 cotton linters, ground and sieved Through a 4 mm screen

by passing all the air through a high-efficiency filter. The gain in mass of this second filter determines the value of C_2.

FILTER PERFORMANCE
Dust Spot Efficiency

Dust spot efficiency or blackness test uses the airborne contaminants already present in the atmosphere. Air is sampled on either side of the filter via target filter papers using equal flow rates. The sample drawn through the downstream target is continuous, whilst that through the upstream is intermittent with a measured time interval such that the opacity of both target papers is comparable. The efficiency is then a function of the total quantity of air passing through each target combined with its light transmission.

Dust Holding Capacity

The dust holding capacity of a filter is the amount of dust that it can hold whilst maintaining its specified efficiency or within its rated pressure drop, from clean to dirty; its value is obtained at the same time as the efficiency tests. For most filters, efficiency and dust loading are interrelated, and therefore, the efficiency is obtained a number of times during the course of a test; the dust holding capacity is the integrated amount of dust measured at each part of the test. A different technique is used for self-renewable filters such as the automatic roll filter.

High-Efficiency Filters

High-efficiency particulate absolute (HEPA) filters are tested using the sodium chloride test that is covered by ISO 29463-1:2011 [4,5]. The test is a discoloration test using small salt crystals that are carried by the airstream. Samples of air upstream and downstream of the filter are passed through a flame of burning hydrogen gas, which becomes bright yellow when exposed to sodium chloride. The intensity of the colour relates directly to the concentration of sodium chloride particles, the brightness of flame being measured with a photosensitive cell.

In the United States, the DOP penetration test is used. In this test, a vapour of dioctyl phthalate is generated with a particle size of approximately 0.3 μm with a cloud concentration of 80 mg/m. Light-scattering techniques are used to obtain the upstream and downstream concentrations.

The performance of high-efficiency filters is often expressed as the *penetration*, *P*, that is given by:

$$P = 100 - \eta \tag{12.2}$$

A DOP standard test is also available for on-site testing.

Face Velocity

The face velocity v_f is the mean velocity of the air entering the effective face area of the filter:

$$v_f = \dot{V}/A_d \tag{12.3}$$

where $\dot{V}$ = volumetric flow rate (m³/s) and A_d = cross-sectional area of duct connection of filter (m²)

Usually, the maximum velocity is recommended by the manufacturer, although this is not necessarily consistent with the face velocity of a filter sized for economic operation.

Pressure Drop

The pressure drop Δp across a filter is related to the face velocity by the following equation:

$$\Delta p = b\,(v_f)^n \tag{12.4}$$

where $1 < n < 2$

For most filters, the value of the constant b will rise depending on the amount of dust it is holding. In much of the literature on filters, the pressure drop is termed, erroneously, the resistance (refer to Chapter 13 for the definition of resistance).

FILTER TYPES

Brief descriptions of the various filters used in air conditioning systems are given below; typical design and operating characteristics are set out in Table 12.2.

Dry Fabric

Dry fabric filters use materials such as cotton wool, glass fibre, cotton fabric, pleated papers, and metal mesh, as the filter medium. The efficiency of the filter depends largely on the area of the medium offered to the airstream. The different types of dry fabric filter include the following:

Table 12.2 Typical design and operating characteristics of air filters

| Filter type | Face velocity (m/s) | Pressure drop | | Efficiency | |
		Initial (Pa)	Final (Pa)	Arrestance A (%)	Dust spot or sodium flame (%)
Panel					
Dry fabric metal mesh	2.0	70	100	70–95	–
Bag					
Low efficiency	2.5	70	300	90–95	45
Medium efficiency	2.0	140	600	98	80–85
High efficiency	1.8	200	600	98	90–95
Automatic roll	2.5	80	160	80–90	–
Absolute					
Low efficiency	2.5	250	500	–	95
Medium efficiency	1.2	250	500	–	99.5
High efficiency	1.2	250	500	–	99.997
Electrostatic	2.5	100	–	90	–

Panel or cell. The filter material is mounted in panels up to 600 mm square. Panels are fitted into a common frame and are placed either at 90 degrees to the air flow or as a series of oblique cells as in Fig. 12.1. The framework is part of either an air handling packaged unit or a separate plant item in the ductwork system. For maintenance, small units are usually side withdrawal, larger units, and front withdrawal; the dirty cells are discarded and replaced with new.

Bag. A set of bags, made of the filter material, is mounted in a frame, with the open ends of the bags facing the airstream as in Fig. 12.2. This provides a large area of material that makes the filter suitable for large volume flow rates at relatively low pressure drops. Filter material is used to provide low, medium, and high efficiencies; the last is often referred to as a semi-absolute filter. The framework and maintenance is similar to that described for panel filters.

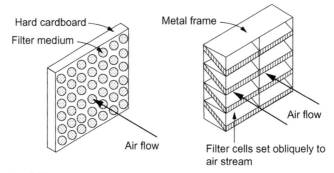

Fig. 12.1 Panel filters.

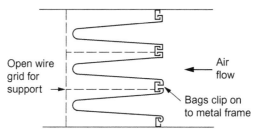

Fig. 12.2 Bag filter (diagrammatic arrangement).

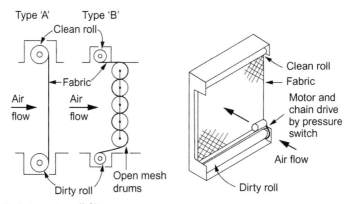

Fig. 12.3 Automatic roll filter.

Automatic roll. In this filter, the medium is stretched between two rollers, one of which is driven by a motor, as in Fig. 12.3. As the filter becomes dirtier, the pressure drop across it increases, and the differential pressure switch starts the motor that rolls on a clean area of filter until the pressure has

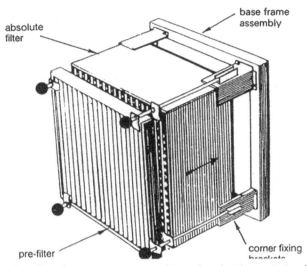

Fig. 12.4 Absolute filter fixing arrangements. *(Reproduced with permission from Mann + Hummel Vokes Air Ltd.)*

dropped to the low limit of the switch. Some media are cleanable, but most are discarded once the whole roll is dirty.

Absolute. HEPA filters use densely packed pleated filter material as shown in Fig. 12.4. Filters are graded under low, medium, and high efficiencies. Some HEPA filters are available to cope with adverse environmental conditions.

The dense filter material results in a high–pressure drop for the rated air flow, and for economic operation, it is often necessary to select a larger filter to give a lower pressure drop, rather than install a unit at the upper limit of at its maximum capacity. A prefilter should be used to remove the relatively large particles from the incoming air and prolong the life of the filter cells; they bolt together to form a multiple unit, and when 'dirty', they are discarded.

Electrostatic Filter

An electrostatic filter has two main sections as shown in Fig. 12.5. The first section is the *ionizing section*; it consists of a series of fine wires charged to a voltage of up to 13 kV, placed alternately with earthed rods. This sets up a corona discharge, and as the airborne particles pass through the ionizing field, they receive a positive electrostatic charge. The second part is the

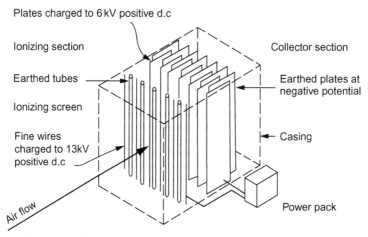

Fig. 12.5 Electrostatic filter.

collector section; this consists of a series of parallel, vertical metal plates with a potential difference of 6–7 kV between adjacent plates. The ionized dust particles are attracted towards these plates to which they adhere. The plates are sometimes coated with oil to help dust retention. The filters are cleaned automatically by washing with high-pressure water.

Adsorption Filter

Adsorption filters are used for the removal of odours, tobacco smoke, and some poisonous gases such as SO_2. The air is passed across a large surface area of activated carbon. The contaminating gas is attracted to the carbon that eventually becomes saturated with the gas. It can be reactivated by removing and heating the carbon to an appropriate temperature.

FILTER SELECTION

The primary requirement is for the filters to remove airborne contaminants at the required efficiency for the application whilst retaining the dust removed from the airstream. Other considerations are as follows:

Environmental conditions. Some conditions, e.g., high humidity and temperature, can adversely affect the filter, and hence, appropriate media should be chosen.

Maintenance. Servicing methods include:
- replacing complete filter;
- renewing filter media;

- reconditioning or cleaning;
- automatic self-cleaning, e.g., electrostatic filters.

Clogged filters can cause failure of systems to supply sufficient air. Suitable indicators and alarms should be provided to ensure filters are maintained; adequate access must be allowed for replacement and cleaning.

Pressure drop. Manufacturers aim to design the filter fabric in such a way as to minimize pressure drop for a given velocity. Nevertheless, the filter should be chosen with as low a pressure drop as possible, consistent with required efficiency and initial costs. This is particularly useful with high-efficiency absolute filters that have high-pressure drops at the maximum recommended face velocities. A derated filter will usually give increased capital costs for builders work and/or ductwork, which is offset by the fan energy savings.

Prefilters. The life of high-efficiency filters can be extended considerably by providing prefilters to remove the larger particles; care should be taken in striking the right economic balance. Prefilters are provided as an integral part of an electrostatic filter system.

Costs. One type of filter may have low initial costs but high operating costs, whereas another type may have high capital cost with low running costs. The initial cost will depend on the filter type, associated ductwork, and any ancillary equipment. The operating cost of the filter is made up of the replacement cost of the media at the end of its useful life, the cost of replacing and/or cleaning the filter media, the energy and cleaning material costs, and the fan power required for the filter pressure drop (the mean value between clean and dirty operation).

A solution to this problem is to calculate the *owner cost-benefit index* (CBI). This index is the ratio of the filter efficiency to the total owning and operating costs. Thus, for a given total cost, the CBI increases with the efficiency and the higher the CBI the more benefit to the client/building owner in terms of

- reduced maintenance, decorating, and housekeeping;
- protection of contents and products;
- protection of high-efficiency (absolute) filters.

SYSTEM DESIGN

Position of the filter in the system. As it is essential to prevent the buildup of contaminants on air conditioning plant items, the filter will normally be the first plant item in the system.

For special areas where high standards of air cleanliness are required, an additional filter may need to be provided as the last plant item. This should be fitted after the fan to cope with any ingress of air, which may occur in the plant. Alternatively, each room outlet may be fitted with a terminal filter.

Recirculated air. A case can be made for cleaning contaminated exhaust air to allow it to be recirculated. The use of potentially contaminated air can be an emotive subject, e.g., recirculation in hospital operating rooms. Even so, techniques for air cleaning are available and may prove to be the most appropriate plant arrangement. The choice will usually depend on the quality of the installation and maintenance to ensure acceptable standards of air cleanliness. An alternative to recirculated air is to use a heat recovery unit.

Changes in air flow rate. The change in air flow rates resulting from increased pressure drop across a dirty filter can be examined by referring to the system and fan characteristics. If the reduction in air flow rate is unacceptable, then near-constant flow rate can be achieved by using one of the following methods:
• Hand reset damper;
• Automatic damper;
• Inlet guide vanes operated with a static pressure controller;
• Variable speed fan.

Protection against fog and frost. Where the filter is to be used in a system employing 100% outdoor air or where a filter is placed in the fresh (outdoor) air duct, fog can sometimes saturate and degrade the filter media; freezing fog can block it completely. Where these conditions are likely, it will be advisable to provide a protective heater upstream of the filter under the control of a low-limit thermostat.

Fresh air intake. The preferred position of the intake is at roof level and sited away from local sources of contamination. Where it is at low level, it should be at least 2 m off the ground; this precaution will reduce the load on the filter. The intake should include a bird and/or insect screen. For buildings that are located in areas liable to sand storms, grilles are fitted to trap any heavy airborne particles.

Installation. Care should be taken to provide adequate seals between the filter units and the holding frames; this is especially important with absolute filters.

REFERENCES

[1] BS EN 779: 2012 Particulate air filters for general ventilation. Determination of the filtration performance, British Standards Institution.

[2] ASHRAE 52-76, Methods of testing air-cleaning devices used in general ventilation for removing particulate matter, 2012.

[3] EUROVENT 4/9–1997, Method of testing air fillers used in general ventilation for determination of fractional efficiency. (Note: There have been various updates to this standard, the latest being EN779–2012.).

[4] BS 3928: 1969, Method for sodium flame test for air filters, British Standards Institution (confirmed 2014).

[5] ISO 29463–1:2011 High-efficiency filters and filter media for removing particles in the air—Part 1: Classification, performance testing and marking.

CHAPTER 13

Fluid Flow: General Principles

In this chapter, the general principles of airflow in ducts are explained, the fluid being treated as incompressible. The relevant equations are given and the general characteristics of flow described. The methods of calculating pressure losses in straight ducts and in fittings are illustrated, together with the pressure distributions. The concept of resistance is explained, and the chapter concludes by giving some relevant methods for measuring flow rates. An understanding of these topics is important before going on in subsequent chapters to consider the design and sizing of ductwork systems, fan selection, and on-site balancing procedures.

Though the emphasis is on air systems, most of the theory is also applicable to piped water systems.

FLOW CONTINUITY

If an incompressible fluid flows steadily in a closed duct (or pipe), the mass flow rate remains constant:

$$\dot{m} = \rho\, A\, \bar{v} = \text{constant} \tag{13.1}$$

If the density remains unchanged, the volume flow rate remains constant:

$$\dot{V} = A\, \bar{v} = \text{constant} \tag{13.2}$$

Example 13.1

The mean air velocity in a 0.4 m diameter duct is 7 m/s, the temperature being 10°C. The duct expands to a rectangular cross section of 0.4×0.8 m to accommodate a heater that raises the air temperature to 50°C Determine the mean velocity of the air in the rectangular duct after the heater.

Solution

Referring to Fig. 13.1:

Air densities corrected for temperature:

circular duct: $\rho = 1.2\, \dfrac{293}{(273 + 10)} = 1.242\,\text{kg/m}^3$

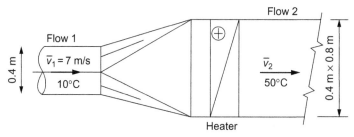

Fig. 13.1 Duct system for Example 13.1.

rectangular duct: $\rho = 1.2 \dfrac{293}{(273+50)} = 1.089 \, \text{kg/m}^3$

Duct areas:

circular duct: $A_1 = \dfrac{\pi \, 0.4^2}{4} = 0.126 \, \text{m}^2$

rectangular duct: $A_2 = 0.4 \times 0.8 = 0.32 \, \text{m}^2$

Using Eq. (13.1):

$$\dot{m} = \rho_1 \, A_1 \, \bar{v}_1 = \rho_2 \, A_2 \, \bar{v}_2$$
$$1.242 \times 0.126 \times 7 = 1.089 \times 0.32 \times \bar{v}_2$$
$$\therefore \bar{v}_2 = 3.14 \, \text{m/s}$$

CONSERVATION OF ENERGY

In any fluid, the total energy is the sum of the potential, pressure, and kinetic energies. (This is summarized by the Bernoulli equation in textbooks on fluid mechanics.) The potential energy, due to the elevation of the fluid above a datum, is assumed constant for an airflow system. Hence, the total energy of the fluid may be expressed in terms of its static (internal) pressure p_s and its velocity (dynamic) pressure p_v. Static pressure acts in all directions at a point in a fluid, whereas velocity pressure can only act in the direction of the flow. This is illustrated in Fig. 13.2.

A pressure gauge A connected flush with the duct wall will read the static pressure, and gauge B connected to a tube facing into the airstream will read the sum of the velocity and static pressures, which is the total pressure, p_t, at the point of measurement, i.e.,

$$p_t = p_v + p_s \tag{13.3}$$

The velocity pressure p_v is given by the relationship:

$$p_v = \tfrac{1}{2} \rho \, v^2 \tag{13.4}$$

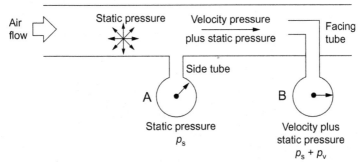

Fig. 13.2 Pressures in a duct.

The pressure relationship of Eq. (13.3) is illustrated by the Pitot-static[1] tube, an instrument used for measuring air pressures in a duct. A diagram of this instrument is shown in Fig. 13.3. The Pitot tube is the inner tube, the head of which is parallel to the duct axis facing directly into the airstream; this measures the total pressure. The static (outer) tube with holes at right angles to the airstream senses only static pressure. By making appropriate connections to a manometer, the total velocity and static pressures can be measured.

Confusion sometimes arises when adding or subtracting pressures because of the difficulty in deciding whether manometer readings are plus or minus. Manometer readings always give the *difference* between two pressures. However, in the case where a Pitot-static tube is used to read velocity pressure, the two pressure readings taken are total and static, and the manometer reads velocity pressure direct; this always has a positive value. Care must be taken to allocate the correct sign to the readings of total and static pressure, which depend on whether the readings are taken on the suction or discharge side of the fan. In each case, using a manometer, the pressure measured is relative to a datum of atmospheric pressure, hence the plus and minus. If absolute pressures are considered (or measured), the confusion does not occur, as may be seen in the following examples.

---

[1] The Pitot tube is named after the French engineer Henri Pitot who invented it in the early 18th century—hence the capital P. It was modified to its modern form in the mid-19th century by the French scientist Henry Darcy.

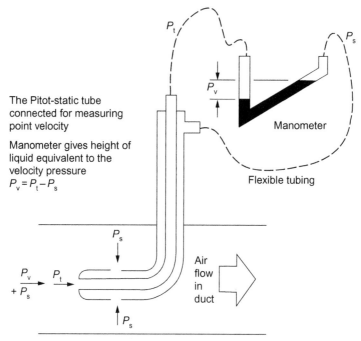

The Pitot-static tube
connected for measuring
point velocity

Manometer gives height of
liquid equivalent to the
velocity pressure
$P_v = P_t - P_s$

P_t

P_v

Manometer

Flexible tubing

P_s

P_v P_t
$+ P_s$

Air
flow
in
duct

P_s

Fig. 13.3 Pressures measured with a Pitot-static tube.

Example 13.2

In a duct on the discharge side of a fan, the total pressure is measured as
100 Pa and the static pressure as 60 Pa, relative to atmospheric pressure
of 1000 mbar. If the air temperature is 30°C, determine the air velocity.

Solution

Referring to Fig. 13.4:

$$\text{Air density} = 1.2 \frac{p_{at}}{1013} \frac{(273 + 20)}{(273 + t_a)} = \frac{1000}{1013} \frac{(273 + 20)}{(273 + 30)}$$

$$= 1.125 \, \text{kg/m}^3$$

Since the pressures are measured on the discharge side of the fan, both are
positive with respect to atmospheric pressure.
Using Eq. (13.3) and remembering that 1 mbar = 100 Pa:
Relative to atmospheric pressure:

$$p_t = p_v + p_s$$
$$100 = 60 + p_v$$
$$\therefore p_v = 40 \, \text{Pa}$$

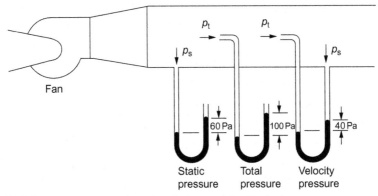

Fig. 13.4 Pressures measured on the discharge side of a fan.

Using Eq. (13.4):

$$p_v = \tfrac{1}{2}\rho v^2$$
$$40 = 0.5 \times 1.125\, v^2$$
$$\therefore v = \sqrt{\frac{40}{0.5 \times 1.125}} = 8.43\,\text{m/s}$$

Example 13.3

On the suction side of the fan in Example 13.2, the total pressure is measured as 50 Pa and the static pressure as 80 Pa, relative to atmospheric pressure of 1000 mbar. Determine the air velocity.

Solution

Referring to Fig. 13.5, since the pressures are measured on the discharge side of the fan, both are positive with respect to atmospheric pressure.

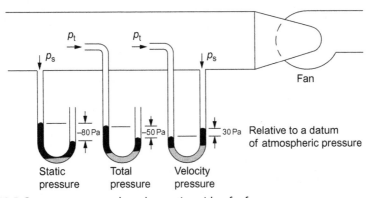

Fig. 13.5 Pressures measured on the suction side of a fan.

Using Eq. (13.3):
Relative to atmospheric pressure:

$$p_t = p_v + p_s$$
$$-50 = -80 + p_v$$
$$\therefore p_v = 30 \, \text{Pa}$$

Using Eq. (13.4):

$$p_v = \tfrac{1}{2}\rho v^2$$
$$30 = 0.5 \times 1.125 \, v^2$$
$$v = \sqrt{\frac{30}{0.5 \times 1.125}} = 7.30 \, \text{m/s}$$

REYNOLDS NUMBER

The ratio of inertia force to viscous force of a fluid flowing in a closed duct or pipe is known as the Reynolds number, Re. For systems to have the same flow regimes at different flow conditions, the Reynolds numbers must be equal:

$$Re_D = \frac{v \, D}{v} \tag{13.5}$$

where v is the kinematic viscosity.

For air at temperature t_a,

$$v = (1.32 + 0.0092 \, t_a) \times 10^{-5} \, \text{m}^2/\text{s} \tag{13.6}$$

Therefore, for air at 20°C:

$$v = 1.5 \times 10^{-5} \, \text{m}^2/\text{s}$$

The diameter D is a characteristic dimension of the duct through which the air is flowing. When the duct is other than circular, the *equivalent* diameter D_e is used; for a rectangular duct, this is known as the *hydraulic* mean diameter, which is given by:

$$D_e = \frac{2A}{(b + W)} \tag{13.7}$$

Example 13.4

Determine the Reynolds number of the air flowing in the rectangular duct in Example 13.1.

Solution

Using the data from Example 13.1:
The kinematic viscosity of air at temperature of 50°C is given by Eq. (13.6):

$$v = (1.32 + 0.0092\, t_a) \times 10^{-5}$$
$$v = (1.32 + 0.0092 \times 50) \times 10^{-5}$$
$$= 1.78 \times 10^{-5}\,\mathrm{m^2/s}$$

The hydraulic diameter is given by Eq. (13.7):

$$D_e = \frac{2A}{(b+W)} = \frac{2 \times 0.4 \times 0.8}{(0.4 + 0.8)} = 0.533\,\mathrm{m}$$

The mean duct velocity in the rectangular duct is 3.14 m/s. The Reynolds number is given by Eq. (13.5):

$$Re_D = \frac{vD}{v} = \frac{3.14 \times 0.533}{1.78 \times 10^{-5}} = 0.94 \times 10^{-5}$$

FLOW CHARACTERISTICS

When air is introduced into a circular, straight duct with a suitably shaped inlet, the velocities near the inlet will be equal at all points in a transverse section up to the vicinity of the duct wall. At this point, the boundary layer is beginning to form, and as the air passes down the duct, wall friction begins to take effect, and the layer becomes thicker. A symmetrical velocity profile develops; beyond a certain distance, the velocity profile remains the same, and this flow is known as *fully developed pipe* flow. The progressive development of the velocity profile along such a duct is shown in Fig. 13.6, though the actual shape of the profile will depend on the Reynolds number, the nature of the flow (which can be either laminar or turbulent), and the roughness of the duct surface.

By introducing a suitable tracer, it can be shown that at low speeds, no mixing occurs but, when the speed has increased beyond a critical value, the tracer becomes rapidly diffused due to small and rapid velocity fluctuations.

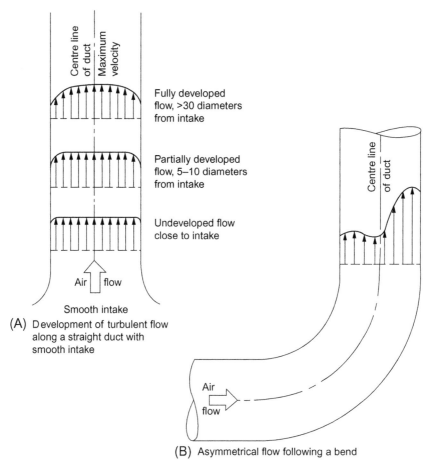

Fig. 13.6 Flow profiles in an air duct.

The flow without mixing is known as *laminar* flow and that with mixing as *turbulent* flow. The transition between the two types of flow occurs at a Reynolds number approximately in the range 2300–4000. Above $Re = 4000$, the flow will be turbulent, and below $Re = 2300$, the flow will be laminar. The velocity at which the change occurs between the two types of flow is known as the *critical velocity*.

The velocities encountered in air ducts are usually well above the critical speed. The lengths of duct required after the entrance section to achieve *fully developed turbulent pipe flow* depend on the intake configuration and the amount of disturbance in the entering fluid. There is no general relationship to predict this length; with a rounded entrance, the distance will be more than 30 diameters of duct length but with an abrupt entry perhaps half this distance.

Asymmetrical velocity profiles, also illustrated in Fig. 13.6, can be produced at the beginning of a straight duct when a certain type of entrance is used, such as a bend or an abrupt constriction. This profile will continue down the duct but with a reducing amount of asymmetry until eventually fully developed flow is established. To achieve this, a straight length of between 50 and 100 equivalent diameters is likely to be required.

The majority of published investigations of fluid flow have been in circular ducts or pipes. Flows in noncircular ducts have not been studied to the same extent though, in the essentials of development and type of flow, it will be similar to that in a circular duct. In addition to the normal flow characteristics, secondary flows are set up, superimposed on the main flow. These secondary flows have the effect of feeding more air into the corner of the duct, thus increasing the velocities in these areas above that which would exist without the effect. However, though not without importance, it is unlikely that these secondary flows have a significant influence on topics of interest to the building services engineer, for example, the accuracy of on-site measurement of flow rate.

PRESSURE LOSSES IN STRAIGHT DUCTS

For fully developed, turbulent flow, the pressure loss due to friction in a straight duct is given by:

$$\Delta p = \frac{fL}{D} \frac{\rho \bar{v}^2}{2} \tag{13.8}$$

or:

$$\Delta p_f = K_f \, p_v \tag{13.9}$$

where:

$$K_f = \frac{f L}{D} \tag{13.10}$$

K_f is termed as the *straight duct loss coefficient*.

The friction factor f varies with duct diameter, the duct wall roughness k, and the Reynolds number. These variations are shown on the Moody chart (Fig. 13.7). The friction factor may be calculated from the following equation:

$$f = \frac{0.25}{\left[\log_{10}(k/3.7D + 5.74/Re_D{}^{0.9})\right]^2} \tag{13.11}$$

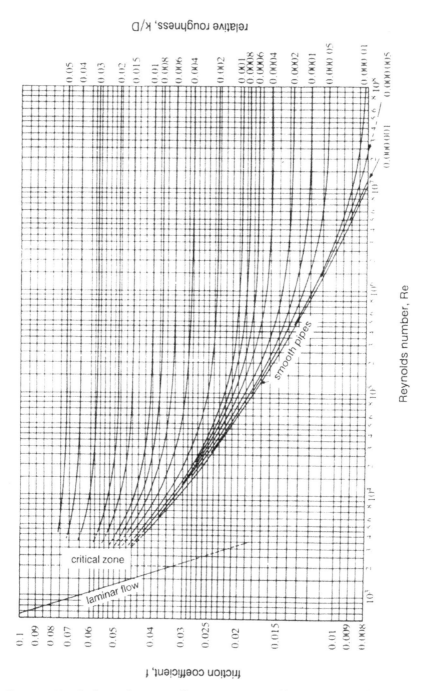

Fig. 13.7 Moody chart—friction coefficient versus Reynolds number. *(Based on Fig. 8.1, D.S. Miller, Internal Flow Systems, BHRA Cranfield, with permission.)*

Example 13.5

A 12 m length of 0.3 m diameter galvanized sheet steel duct carries an air flow rate of 0.4 m³/s. If the average roughness of the duct is 0.15 mm and the air density is 1.2 kg/m³ at 20°C, determine the pressure loss due to friction.

Solution

Duct cross section area: 0.0707 m²
Mean air velocity: 0.4/0.0707 = 5.66 m/s
The Reynolds number is given by Eq. (13.5):

$$Re_D = \frac{vD}{v} = \frac{5.66 \times 0.3}{1.5 \times 10^{-5}} = 1.13 \times 10^{-5}$$

The friction factor is given by Eq. (13.11):

$$f = \frac{0.25}{[\log_{10}(k/3.7D + 5.74/Re_D{}^{0.9})]^2}$$

$$f = \frac{0.25}{[\log_{10}(0.00015/3.70.3 + 5.74/113000^{0.9})]^2}$$

$$f = \frac{0.25}{[\log_{10}(k/3.7D + 5.74/Re_D{}^{0.9})]^2}$$

$$= 0.0201$$

The pressure loss coefficient for the duct is given by Eq. (13.10), i.e.,

$$K_f = \frac{fL}{D} = \frac{0.0201 \times 12}{0.3} = 0.804$$

The velocity pressure is given by Eq. (13.4):

$$p_v = \tfrac{1}{2}\rho v^2 = \frac{1.2 \times 5.66^2}{2} = 19.2 \, \text{Pa}$$

Therefore, the pressure loss due to friction, using Eq. (13.9), is:

$$\Delta p_f = K_f \, p_v = 0.804 \times 19.2 = 15.5 \, \text{Pa}$$

The result for the friction factor in the above example may be confirmed from the Moody chart. The relative roughness, k/D, equals 0.00015/0.3 and at a Reynolds number of 1.13×10^3, the friction factor is 0.0201.

For manual calculations, it is usual to use *a friction chart as* described in Chapter 14, leaving the use of Eqs (13.8) and (13.11) for use in computer programs.

PRESSURE DISTRIBUTION

The total energy in a fluid will remain constant unless changed by external forces. Thus, the total pressure will be constant unless frictional or dynamic pressure losses are present. Therefore, if the velocity changes in a fluid, the static pressure will also change to ensure constant total pressure in a no loss system. Since all components in a system do have pressure losses, the difference in total pressures measured at two points in the system will be equal to the pressure losses between the two points, and the change in static pressure will be determined by using Eq. (13.3). *Relative to a datum of atmospheric pressure*, velocity pressure is always positive; total and static pressures may be either negative or positive depending on whether the duct is on the suction or discharge side of the fan, as explained previously.

Straight Duct Pressure Distribution

With a straight duct of constant cross-sectional area, the velocity pressure remains constant along the length of the duct. Therefore, the static pressure drop along the duct is equal to the pressure loss due to friction.

Example 13.6

Plot the pressure distributions for the 0.3 m diameter duct in Example 13.5. The duct is on the suction side of the fan, and the static pressure at the entrance to the duct is 30 Pa, relative to atmospheric pressure.

Solution

The pressures are plotted in Fig. 13.8.
From Example 13.5, duct velocity $= 5.66$ m/s
    air density $= 1.2$ kg/m^3
    friction pressure drop $= 15.5$ Pa
    velocity pressure $= 19.2$ Pa
Static pressure at end of duct: $p = (-30) - 15.4 = -45.4$ Pa
Using Eq. (13.3), the total pressure at duct entrance:

$$p_{t1} = p_{v1} + p_{s1} = (-30) + 19.2 = -10.8 \, \text{Pa}$$

Total pressure at duct exit:

$$p_{t2} = p_{v2} + p_{s2} = (-45.4) + 19.2 = 26.2 \, \text{Pa}$$

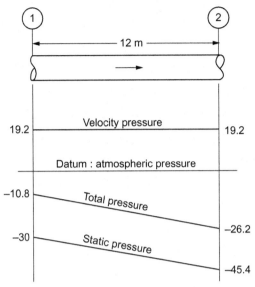

Fig. 13.8 Pressure distribution in a straight duct, Example 13.6.

PRESSURE LOSSES IN DUCT FITTINGS

Fittings in ductwork systems include bends, expansions, contractions, branches, duct inlets from large spaces, discharges to large spaces, and constrictions such as orifice plates. The pressure loss across such fittings is calculated from the product of the velocity pressure p_v and a pressure loss coefficient K, i.e.,

$$\Delta p = K\, p_v \tag{13.12}$$

where p_v is based on the mean duct velocity $\bar{v}$ as determined from Eq. (13.2).

Example 13.7

Determine the pressure loss in a bend that follows the straight length of duct in Example 13.5 and that has a discharge duct of the same cross section. The bend loss coefficient K_b is given as 0.24.

Solution

From Example 13.5:

$$\text{velocity pressure, } p_v = 19.2\,\text{Pa}$$

The pressure loss across the bend is given by Eq. (13.12):

$$\Delta p = K_b\, p_v = 0.24 \times 19.2 = 4.61\,\text{Pa}$$

Determination of Loss Coefficients

Pressure loss coefficients for duct fittings, defined by Eq. (13.12), are obtained from experimental data. In experiments to determine these coefficients, the standard test arrangement is, initially, a straight duct sufficiently long to produce fully developed pipe flow at the upstream face of the fitting; another straight duct downstream of the fitting allows redevelopment of the flow after the disturbance caused by the fitting. The pressure gradient due to friction losses along the duct will be uniform with no fitting present, but with a fitting installed, additional losses occur. These are made up of dynamic and friction losses in the fitting itself and additional friction losses in the downstream duct due to turbulence and velocity gradients generated by the fitting. For calculation purposes, these additional losses are attributed to the fitting, it being assumed that the fitting under test has no duct length. The general relationship between the pressure loss in the straight duct and in a fitting is illustrated in Fig. 13.9.

Tables of loss coefficients for the various fittings, drawn from a number of sources, are given in the following chapter.

Loss coefficients may have some dependence on Reynolds number: the correction factors relating to bends are given in Chapter 14 (Fig. 14.7). However, there is little reliable data for the majority of fittings, and it is unlikely that Reynolds number effects will be used in duct–sizing calculations.

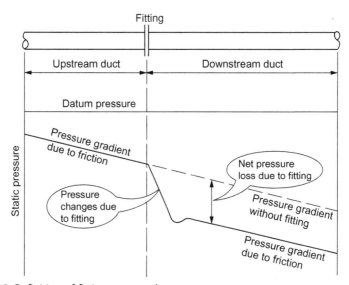

Fig. 13.9 Definition of fitting pressure loss.

Interaction Effects

In ductwork systems installed in buildings, it is often the case that the inlet and outlet duct lengths are too short to allow fully developed pipe flow at the entrance to a fitting and redevelopment of the flow after the disturbance. Fully developed pipe flow described above provides a datum from which departures due to interaction between components can be investigated.

Interaction between two fittings may occur when the length of straight duct between them is insufficient to allow redevelopment of the flow after the first item. Where there is no straight duct between fittings or where the spacer piece is relatively short, the static pressure distributions in the fittings often interact, causing effects far different from those encountered with long lengths of straight duct upstream and downstream of the fitting. Consequently, pressure losses will differ from the sum of the losses in mutually independent fittings.

Miller [1] has investigated and published the interaction effects for a number of typical pairs of fittings, for example, various arrangements of bends, bend/diffuser, and diffuser/bend combinations. Using his data, the net pressure loss across a pair of fittings may be calculated by using a correction factor applied to the sum of their individual loss coefficients based on the inlet velocity pressure to the first component. Some interaction factors for bends are given in Table 14.8.

PRESSURE DISTRIBUTION IN DUCT FITTINGS

Duct Expansion

For a duct expansion, the outlet velocity pressure is lower than inlet velocity pressure. In an *ideal* expansion, there will be no pressure loss, and the total pressure at the outlet will be the equal to that at the inlet. It follows, therefore, that there will be a *rise* in static pressure corresponding to the drop in velocity pressure. Static pressure regain (or pressure recovery) is an important characteristic of an expansion.

Referring to Fig. 13.10 and using Eq. (13.3),

$$p_{t1} = p_{t2} = p_{s1} + p_{v1} = p_{s2} + p_{v2}$$

$$p_{s2} - p_{s1} = p_{v1} - p_{v2}$$

The static pressure regain coefficient R' for maximum pressure recovery is defined by:

$$R'p_{v1} = p_{v1} - p_{v2}$$

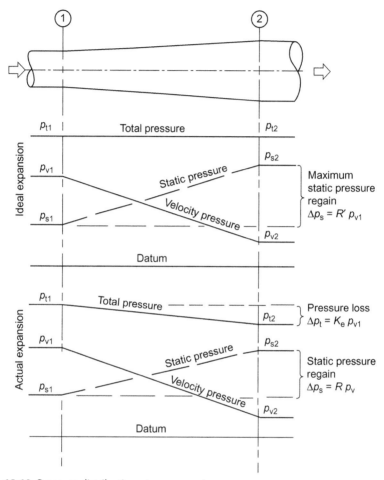

Fig. 13.10 Pressure distribution at an expansion.

$$\therefore R' = 1 - \left(\frac{\bar{v}_2}{\bar{v}_1}\right)^2 \qquad (13.13)$$

For an expansion, the pressure loss is given by Eq. (13.12):

$$\Delta p = K_e\, p_{v1}$$

The static pressure recovery with regain coefficient R is therefore given by:

$$\Delta p = R\, p_{v1} \qquad (13.14)$$

$$R\, p_{v1} = R'\, p_{v1} - K_e\, p_{v1}$$

$$\therefore R = R' - K_e \qquad (13.15)$$

Example 13.8

Plot the pressure distribution for an expansion with an inlet velocity of 10 m/s and an outlet velocity of 5 m/s. The total angle of expansion is 30° and the inlet static pressure 30 Pa relative to the atmosphere on the discharge side of the fan.

Solution

Referring to Fig. 13.11:

Using a standard air density of 1.2 kg/m³, the velocity pressures are:

Inlet: $p_{v1} = \frac{1}{2} \times 1.2 \times 10^2 = 60\,\text{Pa}$

Outlet: $p_{v2} = \frac{1}{2} \times 1.2 \times 5^2 = 15\,\text{Pa}$

For an inclusive expansion angle of 30° (from Table 14.9),

$$K_e = 0.17$$

The total pressure loss across expansion is given by Eq. (13.12):

$$\Delta p_e = K_e\, p_{v1} = 0.17 \times 60 = 10.2\,\text{Pa}$$

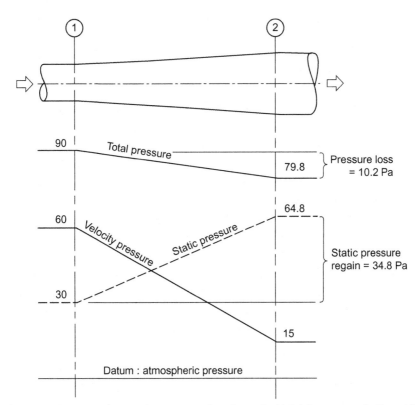

Fig. 13.11 Pressure changes in an expansion, Example 13.8 (all pressures in Pascals).

The pressure recovery coefficient is given by Eq. (13.13):

$$R' = 1 - \left(\frac{\bar{v}_2}{\bar{v}_1}\right)^2 = 1 - 0.5^2 = 0.75$$

The pressure regain coefficient is given by Eq. (13.15):

$$R = R' - K_e = 0.75 - 0.17 = 0.58$$

The static pressure recovery is given by Eq. (13.14):

$$\Delta p = R\, p_{v1} = 0.58 \times 60 = 34.8\,\text{Pa}$$

Static pressure at outlet, $p_s = 30 + 34.8 = 64.8\,\text{Pa}$
Total pressures from Eq. (13.1)
     Inlet: $p_{t1} = p_{s1} + p_{v1} = 30 + 60 = 90\,\text{P}$
     Outlet: $p_{t2} = p_{s2} + p_{v2} = 64.8 + 15 = 79.8\,\text{Pa}$
The difference in total pressures $(90 - 79.8) = 10.2\,\text{Pa}$ is the pressure loss of the expansion.

Contractions

With a contraction, the velocity increases in the downstream duct, and this increase results in a *depression of* the static pressure. For *gradual* contractions, there is only a small pressure loss, and therefore, the static depression is approximately equal to the increase in velocity pressure. Here again, as with a duct expansion, the pressure distribution can be plotted graphically by using the pressure relationship in Eq. (13.3) and illustrated in Fig. 13.12.

System Intake and Discharge

The pressure distribution at an intake to and discharge from a system are illustrated in Figs 13.13 and 13.14.

If there are no losses at the intake, then the static pressure will fall below atmospheric pressure to a value equal to the velocity pressure, thus maintaining zero total pressure. In practise, the static pressure will be lower than the numerical value of p_v by an amount equal to the net pressure loss in the intake. (Use is made of this phenomenon in an inlet to ductwork systems for measuring flow rates, e.g., conical inlets described later in the chapter.)

At the discharge from the system, the total pressure drops to ambient pressure (usually atmospheric pressure) of the space into which the air is being discharged. At the face of the outlet, the system static pressure is at the datum (atmospheric) pressure, and the pressure loss will be the velocity pressure at the outlet.

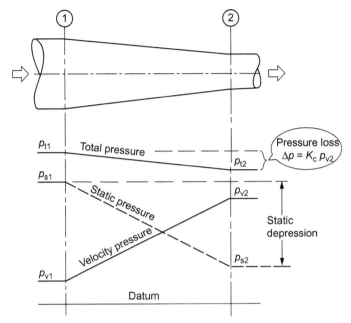

Fig. 13.12 Pressure changes in a contraction.

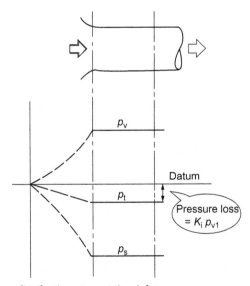

Fig. 13.13 Pressure distribution at a suction inlet.

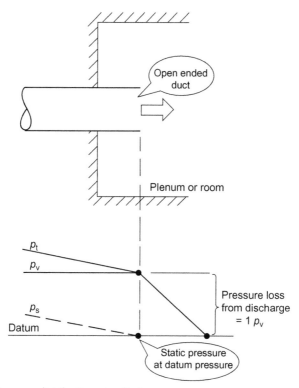

Fig. 13.14 Pressure distribution at a discharge.

RESISTANCE

The concept of resistance is useful in solving various problems in a ductwork system with fully developed turbulent flow. The assumption is that the straight duct friction and the fitting pressure loss coefficients remain constant, that is, they are independent of the Reynolds number. The resistance of a length of a duct, fitting, or plant item in a system is defined by what is known as the *square law*, i.e.,

$$\Delta p = r \, \dot{V}^2 \tag{13.16}$$

Resistance in Series

Three fittings with resistances r_1, r_2, and r_3 connected in series are shown in Fig. 13.15. The flow rate is the same for each item, and the total pressure drop is therefore the sum of the pressure drops across the individual items:

$$\Delta p_t = \Delta p_1 + \Delta p_2 + \Delta p_3$$

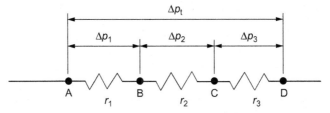

Fig. 13.15 Resistances in series.

Using Eq. (13.16) for each of these pressure drops:

$$r_t\ \dot{V}^2 = r_1\ \dot{V}^2 + r_2\ \dot{V}^2 + \dot{V}^2$$

$$\therefore r_t = r_1 + r_2 + r_3 \tag{13.17}$$

Resistance in Parallel

Three fittings with resistances r_1, r_2, and r_3 connected in parallel are shown in Fig. 13.16. The total flow rate divides into the separate branches, the pressure drop across the nodal points **A** and **B** being the same for each branch:

$$\dot{V}_t = \dot{V}_1 + \dot{V}_2 + \dot{V}_3$$

Using Eq. (13.16) for each of these flow rates:

$$\sqrt{\frac{\Delta p_t}{r_t}} = \sqrt{\frac{\Delta p_1}{r_1}} + \sqrt{\frac{\Delta p_2}{r_2}} + \sqrt{\frac{\Delta p_3}{r_3}}$$

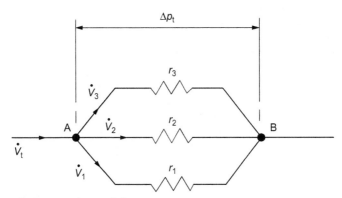

Fig. 13.16 Resistances in parallel.

Since $\Delta p_t = \Delta p_1 = \Delta p_2 = \Delta p_3$, the total resistance is given by:

$$\frac{1}{\sqrt{r_t}} = \frac{1}{\sqrt{r_1}} + \frac{1}{\sqrt{r_2}} + \frac{1}{\sqrt{r_3}} \qquad (13.18)$$

Example 13.9

Part of a system required to handle outdoor air in varying proportions is shown in Fig. 13.17. When operating on 100% outdoor air, the total flow rate is 3 and 1.2 m³/s passing through the preheater. In this mode of operation, the following pressure drops occur:

A to **B**—20 Pa
B to **C**—45 Pa
C to **D**—12 Pa

If at point **A** the pressure at **D** remains constant, determine the change in flow rate through the preheater when the damper is fully shut.

Solution

Referring to Fig. 13.18, which shows resistances in place of the plant items in Fig. 13.17, for the flow rates with the damper fully open, the relevant resistances are calculated using Eq. (13.16):

$$r_1 = \frac{20}{3^2} = 2.22$$

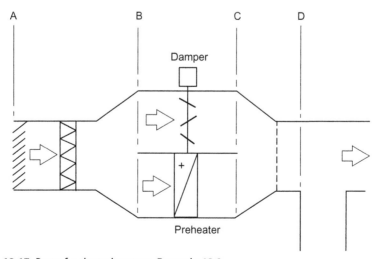

Fig. 13.17 Part of a ducted system, Example 13.9.

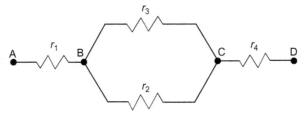

Fig. 13.18 Resistances in parallel, Example 13.9.

$$r_1 = \frac{45}{1.2^2} = 31.2$$

$$r_1 = \frac{12}{3^2} = 1.33$$

With the damper fully shut, the resistances r_1, r_2, and r_3 are in series. Using Eq. (13.17), the total resistance is obtained:

$$r_t = r_1 + r_2 + r_3$$

$$= 2.22 + 31.2 + 1.33 = 34.75$$

The pressure at point $D = 20 + 45 + 12 = 77 \, Pa$
Using Eq. (13.16), the flow rate is given by:

$$\Delta p = r \dot{V}^2$$

$$77 = 34.75 \, \dot{V}^2$$

$$\dot{V} = 1.49 \, m^3/s$$

Therefore, with the damper fully shut, the flow rate through the heater increases by 0.29 m³/s (24%).

Example 13.10

Air is supplied to two outlets that discharge to atmospheric pressure. In the diagram of the system (Fig. 13.19), the pressure at the nodal point **B** is 20 Pa when the flow rates are:

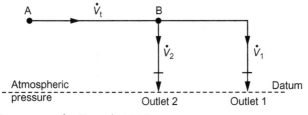

Fig. 13.19 Duct system for Example 13.10.

Outlet 1—0.3 m³/s,
Outlet 2—0.5 m³/s.
If the total flow rate in **AB** is increased to 1.2 m³/s, determine the percentage change in the flow rates from the two outlets.

Solution

Using Eq. (13.16), the resistance are:

Branch 1, $r_1 = \dfrac{20}{0.5^2} = 222$

Branch 2, $r_2 = \dfrac{20}{0.5^2} = 80$

The two branches are resistances, in parallel both flows discharging to atmospheric pressure, which is the datum or zero pressure. Using Eq. (13.18):

$$\frac{1}{\sqrt{r_t}} = \frac{1}{\sqrt{r_1}} + \frac{1}{\sqrt{r_2}}$$

$$\frac{1}{\sqrt{r_t}} = \frac{1}{\sqrt{222}} + \frac{1}{\sqrt{80}}$$

$$\therefore r_t = 31.2$$

The increased pressure at **B** at the increased flow rate of 1.2 m³/s is given by Eq. (13.16):

$$\Delta p = r\,\dot{V}^2$$

$$= 31.2 \times 1.2 = 44.9\,\text{Pa}$$

The flow rate from outlet 1 is given by:

$$\dot{V}_1 = \sqrt{\frac{44.9}{222}} = 0.45\,\text{m}^3/\text{s}$$

The flow rate from outlet 2 is given by:

$$\dot{V}_2 = \sqrt{\frac{44.9}{80}} = 0.75\,\text{m}^3/\text{s}$$

Comparing the increase in flow rates with the original values in Example 13.10, it is seen that both rise by the same percentage. That is, when the total flow rate supplying a pair of outlets is increased (or decreased), the flow rates in the outlets change in the same proportion relative to one another. This is an important principle that is used to develop the on-site proportional balancing procedures explained in Chapter 15.

Pressure Loss Coefficient and Resistance

The relationship between the velocity pressure coefficient and resistance of a fitting or a length of duct is obtained as follows: it is assumed that the fitting has a characteristic area 'A' from which the mean velocity $\bar{v}$ is obtained.

Using resistance, the pressure drop is found from Eq. (13.16):

$$\Delta p = r \, \dot{V}^2$$

Using loss coefficient, the pressure loss is found from Eq. (13.12):

$$\Delta p = K p_v = K \, 0.5 \rho v^2$$

Equating these two relationships:

$$r \dot{V}^2 = K \, 0.5 \rho v^2$$

$$r \dot{V}^2 = K \, 0.5 \rho \left(\frac{\dot{V}}{A}\right)^2$$

$$\therefore r = K \, 0.5 \rho \left(\frac{1}{A}\right) \tag{13.19}$$

That is, resistance is directly proportional to air density.

Example 13.11

A 350 mm diameter bend has a pressure loss coefficient of 0.3. Determine its resistance for an air density of 1.1 kg/m^3.

Solution

$$\text{Cross section area of bend} = \frac{\pi \, 0.35^2}{4} = 0.0962 \, \text{m}^2$$

Using Eq. (13.19):

$$r = K \, 0.5 \rho \left(\frac{1}{A}\right) = 0.3 \times 0.5 \times 1.1 \left(\frac{1}{0.962}\right) = 17.8$$

MEASUREMENT OF FLOW RATE

In the building services industry, there are two basic methods for measuring the flow rate in pipes and ducts:

- By using a pressure drop device with a known flow characteristic;
- By obtaining the mean velocity and multiplying by the duct cross-sectional area.

For commercial and industrial installations, pressure difference devices are used extensively in water and steam pipe distribution networks. Because of the significant additional fan power requirement, large duct cross sections, and inadequate straight lengths of duct, these devices are rarely used for air systems. In the latter case therefore, it is more usual to measure the mean velocity. The principal methods of obtaining the flow rates are described below, using data from the relevant ISO and British standards.

Pressure Difference Devices

Pressure difference devices include orifice plates and conical inlets, and these are illustrated in Fig. 13.20. Other devices include nozzles and venturi metres. A single flow equation is used for these devices, i.e.,

$$\dot{V} = \alpha \varepsilon A_o \sqrt{\frac{2\,\Delta p'}{\rho}} \tag{13.20}$$

Orifice Plates

For orifice plates, the value of the flow coefficient α depends on the orifice to duct diameter ratio β, also to some extent on the duct Reynolds number Re_D and the position of the pressure tappings. Table 13.1 gives typical values of flow coefficients α for Reynolds number of 7×10^4 with pressure tappings placed one duct diameter *upstream* and half a duct diameter *downstream* of the

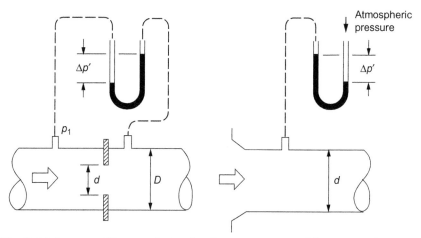

Fig. 13.20 Pressure difference devices for the measurement of flow rate: orifice plate and conical inlet.

Table 13.1 Flow coefficients α for orifice plate with D and D/2 tappings at $Re_D = 7 \times 10^4$

β	α	β	α	β	α
0.20	0.598	0.40	0.610	0.60	0.636
0.22	0.598	0.42	0.614	0.62	0.644
0.24	0.599	0.44	0.616	0.64	0.673
0.26	0.600	0.46	0.619	0.66	0.683
0.28	0.601	0.48	0.623	0.68	0.694
0.30	0.602	0.50	0.627	0.70	0.706
0.32	0.603	0.52	0.631	0.72	0.713
0.34	0.605	0.54	0.636	0.74	0.720
0.36	0.606	0.56	0.642	0.76	0.728
0.38	0.608	0.58	0.649	0.78	0.737

plate [2]. These may be used to determine the flow with reasonable accuracy, but if a more precise calculation is required, then equations, tables, or graphs, given in the relevant standards, may be used. The expansibility factor ε may be taken as unity for the majority of cases dealt with by air conditioning engineers.

Pressure Distribution Through an Orifice Plate

For an orifice plate, mounted in a duct, the air contracts to pass through the orifice and then expands to fill the downstream duct. This expansion of the fluid means that there will be a regain in static pressure and the *net* pressure drop will be less than the pressure difference measured to determine the flow rate. The duct length for full pressure recovery is approximately five equivalent duct diameters. These relative pressure changes are illustrated in Fig. 13.21.

The *net* pressure loss across an orifice plate is related to the measured differential pressure $\Delta p'$ by the following expression:

$$\Delta p = \frac{\left(1 - \beta^2\right)}{\left(1 + \beta^2\right)} \Delta p' \tag{13.21}$$

Conical Inlet

With a conical inlet, as shown in Fig. 13.20B, the differential pressure $\Delta p'$ required for use in Eq. (13.20) is the duct static pressure measured from pressure tappings placed half a duct diameter from the inlet cone. The compound flow coefficient $\alpha\varepsilon$ has some dependence on Reynolds number but may be taken as 0.96 when $Re_D \geq 3 \times 10^5$.

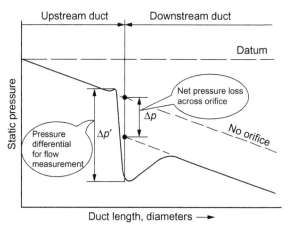

Fig. 13.21 Pressure distribution through an orifice plate.

MEASUREMENT OF MEAN VELOCITY

To obtain the mean velocity, it is usual to traverse the duct with a suitable instrument. The recommended methods for circular and rectangular ducts are detailed below and are independent of the instrument used to measure the point velocities, though it is important that the head of the instrument can be accommodated at the positions close to the duct wall.

In the traverses described, the mean velocity $\bar{v}$ is obtained from the arithmetic average of the total of the individual velocities.

Traverse of a Circular Duct

For a circular duct, the standard traverse is known as a *log-linear traverse*; this has 4, 6, 8, or 10 points per traverse line [3]. The positions for locating the instrument head are given in Table 13.2 and the eight-point traverse illustrated in Fig. 13.22. It is usual to make at least two traverses across the duct at 90° to one another, and the engineer makes a judgement about the number

Table 13.2 Location of measuring points for a log-linear traverse of a circular duct

Points per traverse line	Distance from duct wall, in-duct diameters					
4	0.043	0.290	0.710	0.957		
6	0.032	0.135	0.321	0.679	0.865	0.968
8	0.021	0.117	0.184	0.345		
	0.655	0.816	0.883	0.979		
10	0.019	0.076	0.153	0.217	0.361	
	0.639	0.783	0.847	0.924	0.981	

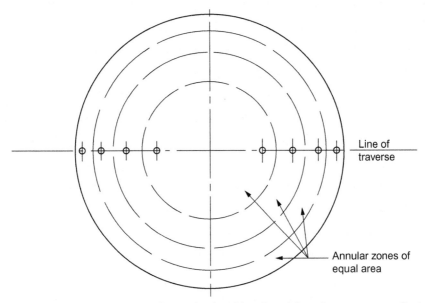

Fig. 13.22 Log-linear traverse of a circular duct (showing eight points per traverse line).

of measuring points. If two eight-point traverses are selected, the mean velocity is obtained from the arithmetic mean of 16 readings of velocity.

Traverse of a Rectangular Duct

The recommended traverse of a rectangular duct is known as a *log-Tchebycheff* (log-*T*) traverse [3,4]. A number (*m*) of straight traverse lines are selected parallel to the smaller side of the rectangle, and on each of them, a number (*n*) of measuring points are located. The positions of the traverse lines and measuring positions are determined in accordance with Table 13.3. Again, the choice of number of measurement points is left to the engineer, with a minimum of 25 and a maximum of 49. A 30-point traverse is illustrated in Fig. 13.23.

Table 13.3 Positions of measuring points and traverse lines for a rectangular duct by the log-Tchebycheff rule

Points per traverse line	Proportional distance of measuring positions from inside wall of duct						
5	0.074	0.288	0.5	0.712	0.926		
6	0.061	0.235	0.437	0.563	0.765	0.939	
7	0.053	0.203	0.366	0.5	0.634	0.797	0.947

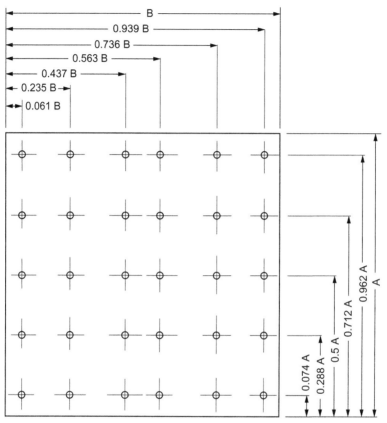

Fig. 13.23 Traverse of a rectangular duct using the log-Tchebycheff rule (five lines traverse lines, each with six points).

INSTRUMENTS FOR MEASURING VELOCITY

The three most used instruments for measuring point velocity are the Pitot-static tube, hot-wire anemometer, and vane anemometer. The application of these instruments to the on-site balancing procedures is described in Chapter 15.

The Pitot-static Tube

The principles behind the Pitot-static tube have already been outlined earlier in the chapter. The point velocity in an airstream is obtained from

$$v = \sqrt{\frac{2\,p_v}{C\rho}} \tag{13.22}$$

where C is the calibration coefficient for the tube.

Values of C are published in ISO 3966:2008 [3]. In determining the flow rate from a Pitot-static tube traverse, additional corrections for effects such as tube blockage and turbulence are applied to the mean velocity. These corrections, amounting to a total of approximately 2% in terms of the mean velocity resulting from a traverse, are usually reserved for laboratory work. For site work, coefficient C may be taken as unity and the other corrections ignored.

Because of its simplicity and ease of use, the Pitot-static tube, connected to a suitable manometer, is used extensively in testing ventilation systems. Normally it is not suitable for use below duct velocities of about 3.5 m/s. Alignment of the probe relative to the flow direction affects accuracy, though this amounts to about only 1% error in the recorded velocity pressure for up to 10° for both yaw and pitch. Most tubes incorporate a short external rod to ensure the probe is aligned to the axis of the duct.

Vane Anemometers

A vane anemometer consists of a number of light vanes supported on radial arms rotating on a common spindle. The main advantage of these instruments compared with a Pitot-static tube is that velocities can be measured down to about 0.5 m/s; the speed range depends on its manufacture, low, medium, and high velocity models being available up to a maximum of about 30 m/s.

A vane anemometer requires a calibration chart, and this should be brought up-to-date regularly to ensure continued accuracy in the measurement of air velocity. The calibration correction v_{corr} is applied to the indicated velocity v_i to give the true velocity v_a. A calibrated instrument has an accuracy of about $\pm 1\%$ over the speed range for which it was designed and with the instrument mounted in a steady, uniform airstream. Yaw error is less than 1% up to 12° of yaw. Variations due to changes in air density are not significant for on-site measurements.

With in-duct measurements, there will be a blockage effect, causing an overestimation of the velocity. Provided the instrument-to-duct diameter ratio is such as to accommodate the instrument at the traverse positions, this effect should not be more than 3%.

These instruments are particularly useful for measurements at the face of a grille, and the recommended procedures for this are outlined in Chapter 16.

Hot-Wire Anemometers

Hot-wire anemometers are highly suited to traversing a duct, and as with a vane anemometer, the main advantage is that velocities can be measured as low as 0.5 m/s, with an accuracy of around ±5%. With the instrument mounted in a steady uniform airstream, the yaw error is less than 5% with up to 15° of yaw.

SYMBOLS

A	duct or pipe cross section
A_o	orifice plate area
b	duct breadth
C	calibration coefficient for Pitot-static tube
D	diameter
D_e	equivalent diameter of rectangular duct
d	diameter of orifice or conical inlet
f	friction factor
K	pressure loss coefficient
K_b	bend pressure loss coefficient
K_e	expansion pressure loss coefficient
K_f	straight duct pressure loss coefficient
k	roughness coefficient of duct wall
L	duct length
$\dot{m}$	mass flow rate
p_{at}	atmospheric pressure
p_s	static pressure
p_t	total pressure
p_v	velocity pressure
Re	Reynolds number
Re_D	Reynolds number related to duct diameter
Re_d	Reynolds number related to orifice diameter
R	pressure regain coefficient
R_s	static pressure regain coefficient
R'	static pressure regain coefficient for maximum pressure recovery
r	resistance
r_t	total resistance
t_a	air temperature
$\dot{V}$	air volume flow rate
v	duct or pipe velocity
$\bar{v}$	mean duct or pipe velocity
W	duct width
α	flow coefficient
β	diameter ratio of orifice to pipe d/D
Δp	pressure drop
Δp_f	pressure loss due to friction
Δp_{rs}	static pressure regain
$\Delta p'$	differential pressure
ε	expansibility factor
ρ	air density
υ	kinematic viscosity

SUBSCRIPTS

1, 2   relates to specific duct or pipe section

REFERENCES

[1] D.S. Miller, Internal Flow Systems, BHRA Fluid Engineering, 1990.
[2] ISO 5167–1: 2003, Measurement of fluid flow by means of pressure differential devices in circular cross-section conduits running full—Part 1:General principles and requirements.
[3] ISO 3966:2008 Measurement of fluid flow in closed conduits—Velocity area method using Pitot-static tubes.

CHAPTER 14

Ducted Air Systems

Ducted air systems can be grouped into two categories with the following general characteristics:

- *Low velocity systems* have air flows of velocities less than 10 m/s; total pressure drop in the ductwork 500 Pa; and plant pressure drops up to 500 Pa;
- *High velocity systems* use air flows in the velocity range 10–40 m/s; pressures on the fan discharge range from 500 to 2500 Pa and on the fan suction up to 500 Pa.

The data presented in this chapter allow the ductwork to be sized and the pressure losses to be determined. Duct sizing procedures are explained and illustrated, and special requirements described.

The design of ductwork distribution systems commences by locating and selecting the room outlet grilles and planning the distribution runs to these outlets in a logical manner, consistent with the building and system design. The layout should attempt to keep the total length of duct to a minimum and the index run as short as is practical. For systems in which the flow rates can be considered to be constant, each section of duct within the network must be sized to carry the sum of the individual outlet flow rates which it serves. For dual duct and VAV systems, account should be taken of the load diversity on the main supply and extract ducts.

When sizing the ducts the engineer should aim at economic sizes, i.e., minimizing the total cost by a comparison of the capital cost of the ductwork and the fan energy cost, whilst at the same time ensuring that air velocities are such that any generated air noise is kept to acceptable levels. With high velocity systems air noise is not usually a problem, since the terminal units will incorporate acoustic attenuating linings.

As the ducts are sized, the pressure losses in the system are calculated, so that the fan pressure requirement can be determined. These pressure losses in air systems may be grouped under three headings:

- friction losses in straight ducts;
- dynamic losses in fittings;
- losses across plant items.

The fan pressure requirement will be the sum of all the losses from these sources along the *index run*.

Air Conditioning System Design
http://dx.doi.org/10.1016/B978-0-08-101123-2.00014-5

PRESSURE LOSSES DUE TO FRICTION

Friction Chart for Circular Ducts

Pressure losses due to friction may be obtained from Eq. (13.8). However, it is more usual to make use of a *duct friction chart,* sometimes referred to as a *duct sizing chart,* as in Fig. 14.1 [1]. This chart has been obtained from the friction equation for the following standard conditions:

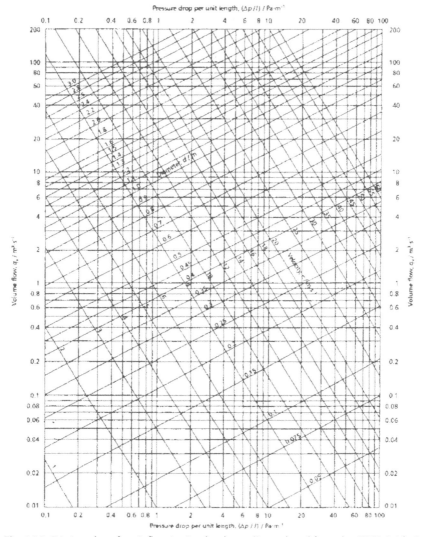

Fig. 14.1 Friction chart for air flow in circular ducts. *(Reproduced from the CIBSE Guide C, by permission of the Chartered Institute of Building Services Engineers.)*

Table 14.1 Recommended sizes of circular duct

Diameter (mm)	Diameter (mm)	Diameter (mm)	Diameter (mm)
63	200	450	800
80	250	500	900
100	315	560	1120
125	355	630	1000
160	400	710	1250

barometric pressure   1013.25 mbar
air temperature   20°C
surface roughness   0.15 mm

The chart is based on the following parameters:

flow rate m^3/s
pressure drop per unit length of duct   Pa/m
duct size   m
mean duct velocity   m/s

When using the chart, interpolation between a pair of lines will often be required, in which case exact values of the parameters may not be obtained. Chart values are sufficiently accurate for most calculations but if a precise value is required, this can be obtained by calculation.

Although ductwork can be made to any size, not all the duct sizes given on the friction chart are *standard* sizes. The recommended standard circular duct sizes are given in Table 14.1 [2].

Example 14.1

A 12 m length of duct is required to supply 0.8 m^3/s of air. Size the duct for a pressure drop of 1.0 Pa/m.

Solution

Refer to Fig. 14.2, a sketch of the friction chart. The intersection of flow rate line 0.8 m^3/s and the pressure drop line of 1 Pa/m gives a duct size of 0.41 m (velocity 6 m/s).

In practise the duct will be sized to the nearest standard diameter size of 400 mm in which case the pressure drop per meter of duct length is (by interpolation, 1.1 Pa).

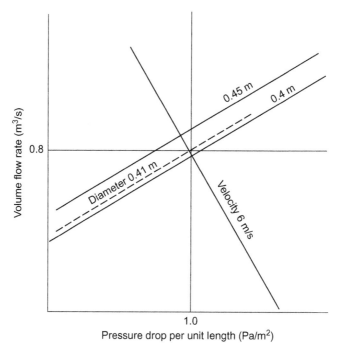

Fig. 14.2 Use of the friction chart.

Rectangular Duct Sizes

Friction charts are available for rectangular ducts but it is more usual to use a conversion table in conjunction with the circular duct friction chart. Equivalent sizes of circular ducts are required so that there will be *equal friction pressure* loss, with *equal surface roughness,* for the *same flow rate.* This will mean that the mean velocity in the rectangular duct will differ from that in the circular duct. The equivalent circular duct size for this requirement is given by:

$$D = 1.265 \left(\frac{(bW)^2}{b+W} \right)^{0.2} \tag{14.1}$$

The rectangular sizes of equivalent circular ducts are given in Table 14.2 [3]. Smaller sizes are read above the stepped line, larger sizes below. The aspect ratio AR of a rectangular duct is defined by the ratio of the dimension of the horizontal side, b, to the vertical side, W, i.e.,

$$AR = \frac{b}{W} \tag{14.2}$$

Table 14.2 Equivalent diameters of rectangular ducts for equal volume flow rate, pressure drop, and surface roughness

Dimension of side of duct, w

h	100	125	150	175	200	225	250	300	350	400	450	500	550	600	650	700	750	800	850	900	950	1000	1050	1100	1150	1200	1250	1300	1400	1500	1600	1700	1800	1900	2000	2100	2200	2300	2400	2500	h
100	110	123	134	145	154	163	171	185	199	211	222	232	242	251	260	268	276	284	291	298																					100
125		138	151	162	173	183	192	209	225	238	251	263	275	285	295	305	314	323	331	339	347																				125
150			165	178	190	201	212	231	248	264	278	291	304	316	327	338	348	358	368	377	386	394																			150
175				193	206	218	230	251	269	287	302	317	331	344	357	369	380	391	401	411	421	430	440																		175
200					220	234	246	269	289	308	325	341	356	371	384	397	409	421	433	444	454	464	474	484																	200
225						248	261	286	308	328	346	364	380	395	410	424	437	450	462	474	485	496	507	517	527																225
250							275	301	325	346	366	385	402	419	434	449	463	477	490	503	515	527	538	549	560	570															250
300								330	357	381	403	424	443	462	479	496	512	527	542	556	570	583	596	608	620	632	643														300
350									385	412	436	459	481	501	520	539	556	573	589	605	620	635	649	662	676	689	701	713													350
400										440	467	492	515	537	558	578	597	616	633	650	667	682	698	713	727	741	755	768	794												400
450											496	522	547	571	594	615	636	655	674	693	710	727	744	760	776	791	805	820	848	874											450
500												551	577	603	627	650	672	693	713	732	751	770	787	804	821	837	853	868	898	927	954										500
550													606	632	658	682	706	728	749	770	790	809	828	846	864	881	898	915	946	976	1005	1033									550
600														661	688	713	738	761	784	806	827	847	867	887	905	923	941	958	992	1024	1054	1084	1112								600
650															716	743	768	793	817	840	862	884	905	925	944	964	982	1000	1035	1069	1101	1132	1162	1191							650
700																771	798	824	849	873	896	918	940	961	982	1002	1022	1041	1077	1112	1146	1179	1210	1240	1269						700
750																	826	853	879	904	928	952	974	997	1018	1039	1059	1079	1118	1154	1189	1223	1256	1287	1318	1347					750
800																		881	908	934	959	984	1007	1030	1053	1075	1096	1116	1156	1195	1231	1266	1300	1333	1365	1396	1425				800
850																			936	963	989	1015	1039	1063	1086	1109	1131	1152	1194	1234	1272	1308	1343	1378	1410	1442	1473	1503			850
900																				991	1018	1044	1070	1095	1119	1142	1165	1187	1230	1271	1311	1349	1385	1420	1455	1488	1520	1551	1581		900
950																					1046	1073	1100	1125	1150	1174	1198	1221	1265	1308	1349	1388	1426	1462	1497	1532	1565	1597	1629	1659	950
1000																						1101	1128	1155	1180	1205	1230	1253	1299	1343	1385	1426	1465	1503	1539	1574	1609	1642	1675	1706	1000
1050																							1156	1183	1210	1236	1261	1285	1332	1378	1421	1463	1503	1542	1580	1616	1651	1686	1719	1752	1050
1100																								1211	1238	1265	1291	1316	1365	1411	1456	1499	1540	1580	1619	1657	1693	1728	1763	1796	1100
1150																									1266	1294	1320	1346	1396	1444	1490	1534	1576	1618	1657	1696	1733	1770	1805	1840	1150
1200																										1321	1349	1375	1426	1476	1523	1568	1612	1654	1695	1735	1773	1810	1847	1882	1200
1250																											1376	1404	1456	1507	1555	1601	1646	1690	1732	1772	1812	1850	1887	1924	1250
1300																												1432	1485	1537	1586	1634	1680	1724	1767	1809	1850	1889	1927	1964	1300
1400																													1542	1596	1647	1697	1745	1792	1837	1880	1923	1964	2004	2043	1400
1500																														1652	1706	1758	1808	1856	1903	1949	1993	2036	2078	2119	1500
1600																															1762	1816	1868	1919	1967	2015	2061	2106	2149	2191	1600
1700																																1872	1926	1978	2029	2078	2126	2172	2218	2262	1700
1800																																	1982	2036	2089	2140	2189	2237	2284	2330	1800
1900																																		2092	2146	2199	2250	2300	2348	2395	1900
2000																																			2202	2257	2309	2361	2411	2459	2000
2100																																				2312	2367	2420	2471	2521	2100
2200																																					2423	2477	2530	2581	2200
2300																																						2533	2587	2640	2300
2400																																							2643	2697	2400
2500																																								2753	2500

Dimension of side of duct, w

$$d_e = \frac{1.453\,A^{0.6}}{P^{0.2}}$$

(Reproduced from the CIBSE Guide C, by permission of the Chartered Institute of Building Services Engineers.)

Example 14.2

A rectangular duct with dimensions 500 and 1200 mm is to supply $4.0 \text{ m}^3/\text{s}$ of air at 20°C to a large hall. Determine the pressure drop due to friction if the duct is 30 m in length.

Solution

Refer to the sketch of Table 14.2 in Fig. 14.3:

Duct diameter $= 837 \text{ mm}$ or 0.84 m

From the friction chart, Fig. 14.1:

0.84 m diameter duct with a flow rate of $4.0 \text{ m}^3/\text{s}$ pressure drop per unit Length $= 0.72 \text{ Pa/m}$

$$\therefore \text{ pressure drop in duct} = 30 \times 0.72 = 21.6 \text{ Pa}$$

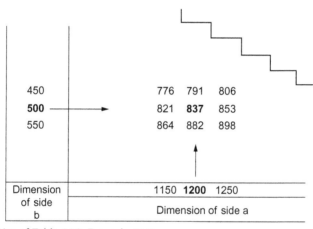

450		776	791	806
500	→	821	**837**	853
550		864	882	898

Dimension of side b	1150 **1200** 1250
	Dimension of side a

Fig. 14.3 Use of Table 14.2: Example 14.2.

Corrections for Air at Other Densities

The pressure losses obtained from the friction chart may be corrected for other air densities by using the following expression:

$$\Delta p_2 = \Delta p_1 \left(\frac{\rho_2}{\rho_1}\right) \tag{14.3}$$

where $\Delta p_1 =$ pressure loss at standard density ρ_1 (1.2 kg/m^3) and $\Delta p_2 =$ pressure loss at air density ρ_2.

Corrections for Ducts of Other Materials

The friction chart is based on an average roughness coefficient of 0.15 mm, assumed to be typical of commercial galvanized sheet steel ductwork.

Table 14.3 Pressure drop correction factors for ducts of different materials (galvanized sheet steel = 1.0)

Pressure drop from chart (Pa/m)	Spiral wound	Plastic	Builders work			
			Fair-faced brick or concrete		Rough brickwork	
			200 mm	1000 mm	200 mm	100 mm
0.5	0.95	0.88	1.42	1.42	2.18	2.05
1.0	0.94	0.85	1.48	1.46	2.34	2.12
2.0	0.93	0.82	1.53	1.49	2.45	2.17

Note: the dimensions are *equivalent* diameters.

For ducts of other materials, with different roughness values, the pressure drops taken from the chart can be corrected with the factors given in Table 14.3.

PRESSURE LOSSES IN FITTINGS

The general principles of determining the pressure loss across duct fittings are given in Chapter 13. The specific loss coefficients for different types of fittings, drawn in the main from Miller [4] and the ASHRAE Handbook [5], are given below together with examples to illustrate calculation methods. There are some variations in the loss coefficients quoted by these sources; the values given in the following tables and graphs are considered to be those which best reflect the probable loss. In selecting this data priority has been given to the fittings recommended in the BESA Specification for Sheet Metal ductwork [6].

It will be appreciated that, where more than one velocity is associated with a fitting, the loss coefficient must be applied to the appropriate velocity pressure. For example, in the case of expansions and contractions, the velocity pressure in the smaller area is used to calculate the loss; for branch pieces, the velocity pressure in the duct carrying the total flow. (Other texts may use different reference velocities.)

In general, pressure losses in fittings apply to both supply and extract systems, the main exception being those for branch pieces.

Bends

Circular Ducts

The loss coefficients for circular duct bends are given in Table 14.4. The following bend construction details are recommended:

- for ducts up to, and including, 300 mm: long radius pressed bends, $r/D = 1.5$;

Table 14.4 Loss coefficients K_b for circular duct bends

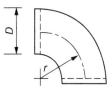

90 degrees radiused pressed bend

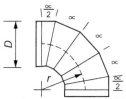

90 degrees segmented bend

	Radius to duct diameter ratio, r/D					Segmented bends	
Bend angle	– 0.75	Medium radius 1.0	Long radius 1.5	– 2.0	– 2.5	No. of segments	α
90 degrees	0.40	0.24	0.18	0.16	0.15	5	22.5
60 degrees	0.22	0.15	0.12	0.11	0.11	4	20
45 degrees	0.14	0.10	0.09	0.09	0.09	3	15
30 degrees	0.07	0.06	0.06	0.06	0.06	2	15

- for ducts up to, and including, 400 mm: medium radius pressed bend, $r/D = 1.0$;
- for ducts above 400 mm: segmented bends, see Table 14.5.

Example 14.3

Determine the pressure loss across a medium radius, 90 degrees bend for a 350 mm diameter circular duct. The air flow rate is 0.6 m³/s at standard air density.

Solution

The mean velocity is given by:

$$\bar{v} = \frac{\dot{V}}{A} = \frac{0.6}{\pi\,0.35^2/4} = 6.24\,\mathrm{m/s}$$

The velocity pressure is given by:

$$p = 0.6\,\bar{v}^2 = 0.6 \times 6.24^2 = 23.3\,\mathrm{Pa}$$

For a medium radius bend $r/D = 1.0$

From Table 14.4 the loss coefficient is 0.24

From Eq. (12.12), the pressure loss across bend is obtained:

$$\Delta p_b = K_b p_v = 0.24 \times 23.3 = 5.6\,\mathrm{Pa}$$

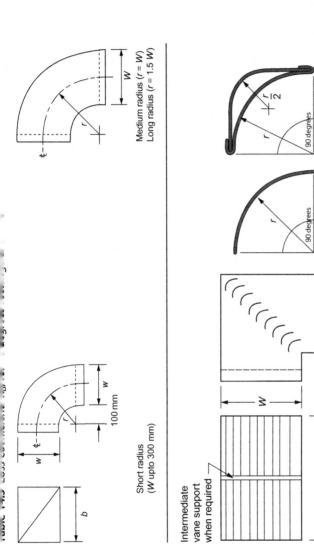

Short radius
(W upto 300 mm)

Medium radius (r = W)
Long radius (r = 1.5 W)

Intermediate
vane support
when required

Mitred bend

Vanes for mitred bends

Aspect ratio, b/W

Bend description	$\dfrac{r}{W}$	Hard bends				Easy bends		
		0.25	0.5	1.0	2.0	3.0	4.0
Short radius	0.8	0.57	0.50	0.44	0.40	0.40	0.40
Medium radius	1.0	0.27	0.25	0.21	0.18	0.18	0.19
Long radius	1.5	0.22	0.20	0.17	0.14	0.14	0.15

Notes: (a) dimensions apply to other turning angles; (b) multipliers for bend angles less than 90 degrees: 60 degrees—0.8, 45 degrees—0.6, 30 degrees—0.3; (c) 90 degrees mitred bends with no turning vanes $K_b = 1.1$. 90 degrees mitred bends with turning vanes $K_b = 0.15$.

Rectangular Ducts

Bends for rectangular ducts are described as *hard* or *easy*. A hard bend rotates in the plane of the longer side and an easy bend rotates in the plane of the shorter side; these are illustrated in Fig. 14.4. The loss coefficients are given in Tables 14.5 and 14.6. The following bend construction details are recommended:

- for ducts up to 300 mm in duct width: short radius bend, throat radius $= 100$ mm;
- for ducts over 300 mm in duct width: short radius bend with splitters as specified in Table 14.6;
- for all duct widths: medium radius bends, $r/W = 1.0$ long radius bends, $r/W = 1.5$.

Example 14.4

Determine the pressure loss across a long radius, 90 degrees bend for a rectangular duct width 200 mm and height 400 mm. The air flow rate is 0.6 m^3/s at standard air density.

Solution

The mean velocity is:

$$\bar{v} = \frac{0.6}{0.4 \times 0.2} = 7.5 \, m/s$$

The velocity pressure is:

$$p_v = 0.6 \, \bar{v}^2 = 0.6 \times 7.5^2 = 33.7 \, Pa$$

For a long radius bend $r/W = 1.5$

The aspect ratio $\dfrac{b}{W} = \dfrac{200}{400} = 0.5$

From Table 14.5 the loss coefficient $= 0.20$

Therefore the pressure loss across bend is obtained:

$$(\Delta p = K_b p_v = 0.20 \times 33.7 = 6.7 \, Pa)$$

The loss coefficient for a *mitred bend* can be taken as 1.1. By including turning vanes (see Table 14.5) the bend loss will be reduced to that of the lowest value of an equivalent arc bend. The vanes, single or double skinned, usually have an inner radius r of 50 mm and the recommended number of vanes n is given by the relationship:

$$n = \frac{1.5 \, W}{r} \tag{14.4}$$

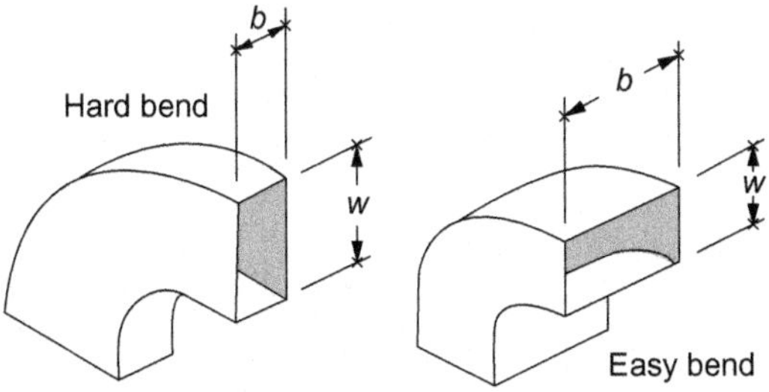

Fig. 14.4 Hard and easy rectangular bends.

Table 14.6 Loss coefficients K_b for 90 degrees rectangular duct bends with splitters

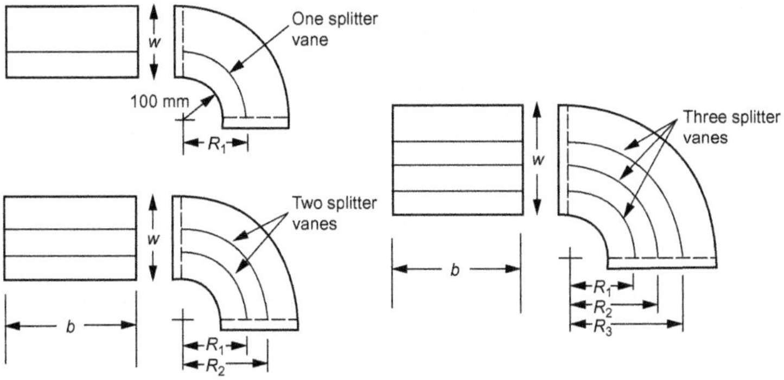

90 degrees short radius bends (*W* over 300 mm)
$R_1 = 100/C_r$, $R_2 = 100/C_r^2$, $R_3 = 100/C_r^3$
where C_r = radius curve ratio for splitter vane

No. of splitters	Duct width *W*	Radius curve ratio C_r	Aspect ratio, *b/W*					
			Hard bends		–	Easy bends		
			0.25	0.5	1.0	2.0	3.0	4.0
1	300	0.50	0.13	0.09	0.08	0.07	0.08	0.08
	400	0.45	0.18	0.13	0.11	0.11	0.12	0.13
	500	0.41	0.22	0.16	0.14	0.15	0.16	0.17
2	500	0.55	0.09	0.07	0.06	0.06	0.06	0.06
	700	0.50	0.12	0.09	0.08	0.08	0.09	0.10
	1000	0.45	0.17	0.13	0.11	0.13	0.15	0.16
3	1000	0.55	0.07	0.05	0.06	0.06	0.07	0.07
	2000	0.47	0.11	0.10	0.12	0.14	0.16	0.18

Notes: (a) dimensions apply to other turning angles; (b) multipliers for bend angles less than 90 degrees: 60 degrees—0.8, 45 degrees—0.6; (c) splitters are not applicable to bend angles less than 45 degrees.

Corrections for Reynolds Number

Pressure loss coefficients have some dependence on REYNOLDs number, although the data available to allow for its effect is limited. The data for bends is included here to illustrate how a basic loss coefficient is modified by a REYNOLDs number correction factor, C_{Re}, applied to the basic loss coefficient, K_b [7]. The variations of C_{Re} are given in Fig. 13.5 and the pressure loss is calculated from the following equation:

$$\Delta p = C_{Re}K_b p_v \qquad (14.5)$$

Example 14.5

For the bend in Example 14.4, determine the Reynolds number correction factor to be applied to the basic loss coefficient.

Solution

For the bend given, hydraulic diameter is given by Eq. (13.7):

$$D_e = \frac{2A}{(b+W)} = \frac{2(0.2 \times 0.4)}{(0.2+0.4)} = 0.267$$

Reynolds number:

$$Re = \frac{D\bar{v}}{v} = \frac{0.267 \times 7.5}{1.5 \times 10^{-5}} = 1.33 \times 10^5$$

From Fig. 14.5, the Reynolds number correction factor for r/D of 1.5:

$$C_{Re} = 1.3$$

Therefore, using Eq. (14.5), the pressure loss of the bend is:

$$\Delta p = C_{Re}K_b p_v$$
$$= 1.3 \times 0.24 \times 23.3 = 7.3 \text{Pa}$$

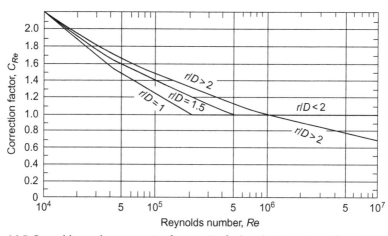

Fig. 14.5 Reynolds number correction factors, C_{Re}, for bends. (Reproduced from D.S. Miller, Internal Flow Systems, BHRA Cranfield, with permission.)

Table 14.7 Typical bend interaction effect factors C_{b-b}

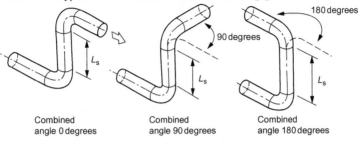

| Combined | Combined | Combined |
| Combined angle 0 degrees | Combined angle 90 degrees | Combined angle 180 degrees |

	Spacer length, duct diameters, L_s			
Combined bend angle	**0**	**1**	**4**	**8**
0 degrees	1.00	0.86	0.71	0.81
90 degrees	0.81	0.79	0.74	0.82
180 degrees	0.53	0.58	0.71	0.80

Correction for Interaction Between Bends

When bends follow each other there is an interaction affecting the total pressure loss of the system and this applies to ductwork in which the intervening (spacer) length is less than 30 diameters. To arrive at the net loss of a pair of bends, a correction factor C_{b-b} is applied to the sum of the individual loss coefficients. Then the pressure loss is given by:

$$\Delta p = C_{b-b}(K_{b1} + K_{b2})p_v \qquad (14.6)$$

A typical set of interaction effect factors are given in Table 14.7 for circular duct bends with an r/D ratio of 1.0. The bend configurations referred to in this table are defined by the sketches. The factors assume that the outlet discharge duct from the bend arrangement is more than 30 diameters in length. This data can also be used for rectangular duct bends but where the bends incorporate turning vanes there is no interaction effect.

Example 14.6

Two 90 degrees medium radius, 400 mm diameter bends are separated by 1.6 m of straight duct and are arranged with a combination angle of 180 degrees. Determine the net pressure loss due to the bends if the flow rate is 1.2 m³/s, at standard air density.

Solution

Mean duct velocity:

$$\bar{v} = \frac{\dot{V}}{A} = \frac{1.2}{\pi 0.4^2/4} = 9.55 \, \text{m/s}$$

Velocity pressure:
$$p_v = 0.6 \times 9.55^2 = 54.7 \, \text{Pa}$$
From Table 14.4: $K_{b1} = K_{b2} = 0.24$
The length of the spacer in duct diameters is given by:
$$L_s = 1.6/0.4 = 4$$
From Table 14.7, the interaction correction factor $C_{b-b} = 0.71$
Using Eq. (14.6):
$$\Delta p = C_{b-b} \left(K_{b1} + K_{b2} \right) p_v$$
$$= 0.71 \left(0.24 + 0.24 \right) 54.7 = 18.6 \, \text{Pa}$$

Expansions

Ducts which expand in cross-sectional area are variously termed expansions, expanders, diffusers, and enlargements. Expansions are classified into two groups:
- Ducted discharge;
- No discharge duct (i.e., a free discharge).

Pressure loss coefficients for these two groups of fittings are given in Tables 14.8 and 14.9. The coefficients are for both circular and rectangular ducts and are applied to the inlet velocity pressure.

For circular ducts it is recommended that the slope angle of an expansion should not exceed 15 degrees on any side and for rectangular ducts, 22.5 degrees.

For rectangular ducts, splitters (vanes) may be used to minimize the pressure loss. For vaned expanders with a free discharge, a reasonable performance will be achieved with a half angle:

θ up to 20 degrees—two vanes
θ up to 25 degrees—three vanes

The recommended angle θ of the expansion vanes, from the axis is given by:

$$\alpha = \frac{\theta}{2.9}$$

where $\alpha = $ angle of expansion (see figure in Table 14.9)

Expansion With a Short Outlet Pipe

When an outlet duct is added to an expansion, there is a reduction in the loss coefficient related to the length of the outlet duct. The loss coefficient, K_{es}, is obtained from the following relationship:

$$K_{es} = 1 - C_e \left(1 - K_e \right) \tag{14.7}$$

where K_{es} is obtained from Table 14.7 and C_e is the loss correction factor in Table 14.10.

Table 14.8 Loss coefficient K_e for circular and rectangular duct expansions with long discharge duct

Area ratio A_2/A_1	Velocity ratio v_2/v_1	Half-angle, θ			
		5 degrees	10 degrees	15 degrees	22.5 degrees
1.2	0.83	0.06	0.06	0.06	0.06
1.5	0.67	0.06	0.08	0.09	0.11
2.0	0.50	0.06	0.12	0.17	0.21
2.5	0.40	0.08	0.17	0.20	0.30
3.0	0.33	0.10	0.20	0.30	0.40
3.5	0.29	0.20	0.35	0.45	0.50
4.0	0.25	0.27	0.42	0.50	—

Note: maximum angles: circular duct $\theta = 15$ degrees rectangular duct $\theta = 22.5$ degrees.

Table 14.9 Loss coefficient K_e for circular and rectangular expansions with free discharge

Description	Area ratio A_2/A_1	Velocity ratio $\bar{v}_2/\bar{v}_1$	Half-angle, θ			
			5 degrees	10 degrees	15 degrees	20 degrees
Circular, concentric	1.5	0.67	0.57	0.62	0.70	—
	2.0	0.50	0.44	0.57	0.70	—
	2.5	0.40	0.37	0.55	0.65	—
	3.0	0.33	0.32	0.50	0.65	—
	3.5	0.29	0.30	0.55	0.70	—
	4.0	0.25	0.30	0.58	0.70	—
Rectangular, concentric	1.5	0.67	0.52	0.60	0.70	0.75
	2.0	0.50	0.37	0.55	0.65	0.75
	2.5	0.40	0.30	0.48	0.65	0.75
	3.0	0.33	0.30	0.45	0.65	0.80
	3.5	0.29	0.30	0.45	0.65	0.80
	4.0	0.25	0.30	0.45	0.65	0.80
Rectangular, concentric with splitters	2.0	0.50	—	0.45	0.45	0.45
	3.0	0.33				
	4.0	0.25	—			

Note: coefficients include the discharge velocity pressure.

Table 14.10 Loss correction factors C_e for short outlet duct (length n) following an expansion

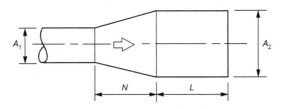

L/D$_2$	Area ratio, A$_2$/A$_1$	Velocity ratio, v$_2$/v$_1$	N/D$_1$ 1	N/D$_1$ 2	N/D$_1$ 3
1	1.5	0.67	1.04	1.00	1.00
	2.0	0.50	1.15	1.08	1.05
	2.5	0.40	1.25	1.13	1.08
	3.0	0.33	1.30	1.17	1.08
2	1.5	0.67	1.07	1.03	1.00
	2.0	0.50	1.28	1.13	1.08
	2.5	0.40	1.50	1.24	1.16
	3.0	0.33	1.60	1.32	1.15
4	1.5	0.67	1.10	1.00	1.00
	2.0	0.50	1.45	1.20	1.08
	2.5	0.40	–	1.40	1.22
	3.0	0.33	–	1.53	1.24

Note: the factors are applied to the loss coefficients K_b in Table 14.7, using Eq. (14.7).

Example 14.7

An expansion in a circular duct with inlet diameter 0.2 m, discharge duct diameter 0.3 m, and angle of expansion on both sides of 10 degrees. If the air flow rate is 0.2 m³/s at an air density of 1.15 kg/m³, determine the pressure loss for:

(a) with long discharge duct;
(b) with no discharge duct;
(c) with discharge duct 0.6 m in length.

Solution

Inlet duct area $A_1 = \pi 0.2^2/4 = 0.0314\,\text{m}^2$
Outlet duct area $A_2 = \pi 0.3^2/4 = 0.0707\,\text{m}^2$
The velocity at the inlet is:

$$\bar{v}_1 = \frac{\dot{V}}{A_1} = 0.2/0.0314 = 6.37\,\text{m/s}$$

The inlet velocity pressure is:

$$p_{v1} = 0.5\rho\,\bar{v}_i^2 = 0.5 \times 1.15 \times 6.37^2 = 23.3\,\text{Pa}$$

$$\text{The area ratio} = \frac{A_2}{A_1} = \frac{0.0707}{0.0314} = 2.25$$

(a) From Table 14.8, the loss coefficient $K_e = 0.145$ (by interpolation)
The pressure loss across expansion with *long outlet disgorge* duct is:

$$\Delta p = K_e p_v = 0.145 \times 23.3 = 3.4\,\text{Pa}$$

(b) From Table 14.9, the loss coefficient $K_e = 0.56$
$\therefore$ the pressure loss across expansion with *free discharge* is:

$$\Delta p = K_e p_v = 0.56 \times 23.3 = 13.0\,\text{Pa}$$

(c) $N = \dfrac{0.05}{\tan^\circ} = 0.284\,\text{m}$

$$\therefore N/D_1 = 0.284/0.2 = 1.42$$

$$L/D_2 = 0.6/0.3 = 2$$

From Table 14.9, the loss factor $C_e = 1.18$ (by interpolation)
Using Eq. (14.7):

$$K_{es} = 1 - C_e(1 - K_e)$$

$$= 1 - 1.18\,(1 - 0.56) = 0.45$$

$\therefore$ the pressure loss across expansion with a *short outlet* duct is:

$$\Delta p = K_{es} p_v = 0.45 \times 23.3 = 10.5\,\text{Pa}$$

Contractions

Ducts which reduce in cross-sectional area are either called contractions or reducers.

Pressure loss coefficients K_c for abrupt contractions, given in Table 14.11, are applied to the velocity pressure in the outlet duct. The recommended maximum slopes of gradual contraction pieces are given in Table 14.12. There is a single loss coefficient K_c of 0.05.

Example 14.8

Determine the pressure loss across an abrupt contraction with an inlet area of $0.4\,\text{m}^2$ and an outlet area of $0.15\,\text{m}^2$. The air flow rate is $0.9\,\text{m}^3/\text{s}$ at a standard air density of $1.2\,\text{kg/m}^3$.

Solution

The mean velocity at the outlet is:

$$\bar{v} = \dot{V}_2/A = 0.9/0.15 = 6.0 \text{m/s}$$

The velocity pressure is:

$$P_{v2} = 0.5\rho\bar{v}^2 = 0.5 \times 1.2 \times 6.0^2 = 21.6 \text{Pa}$$

The area ratio is:

$$A_2/A_1 = 0.15/0.4 = 0.38$$

From Table 14.11 the loss factor for an abrupt contraction $K_c = 0.46$
∴ the pressure loss across the abrupt contraction is:

$$\Delta p = K_c p_v = 0.46 \times 21.6 = 9.9 \text{Pa}$$

Table 14.11 Loss coefficients K_c for abrupt contractions

Area ratio, A_2/A_1	K_C
0.2	0.55
0.3	0.48
0.4	0.45
0.5	0.38
0.6	0.30
0.7	0.20
0.8	0.08

Table 14.12 Loss coefficient K_c for gradual contractions

Contraction	Maximum values of angle, θ	
	Rectangular	Circular
Concentric	22.5 degrees	15 degrees
	0.05	
Eccentric	22.5 degrees	30 degrees
	0.05	

Branches

The alternative arrangements for making a branch connection are shown in Figs 14.6–14.8. These apply to both supply and extract systems. It is recommended that a branch should be taken off the main duct and not off a taper. Therefore, when a change of section is required to accommodate the next section of ductwork, expansion or contraction may be necessary as illustrated in Fig. 14.9.

For supply ducts, the loss coefficients K_{31} and K_{32} refer to the velocity pressure in the duct with the *combined flow*, and this duct is designated as duct 3:

K_{31} refers to the loss to branch 1

K_{32} refers to the loss to branch 2, (usually the main duct)

The loss coefficients K_{31} for 45 degrees and 90 degrees square branch connections for circular ducts are given in Fig. 14.11. These figures give the contours of constant loss coefficients and are typical of data given by Miller [8].

For conical and shoe branch connections, correction factors C_{31} in Table 14.13 should be applied to the square branch coefficient K_{31}, i.e.,

$$\Delta p = C_{31} K_{31} p_{v3} \tag{14.8}$$

Loss coefficients K_{32} for the pressure loss in the main are given in Table 14.14; these are independent of any chamfers or branch angles.

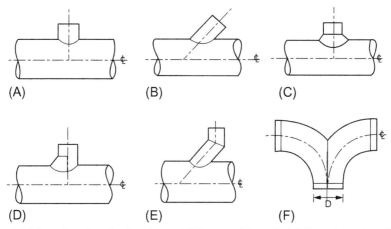

Fig. 14.6 Branches for circular ducts. (A) Square, (B) angled (all angles other than 90 degrees), (C) conical, (D) shoe, (E) mitred, and (F) tee.

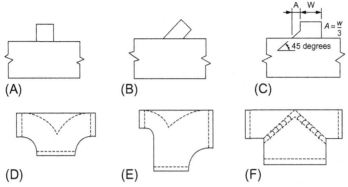

Fig. 14.7 Branches for rectangular ducts [branch types (A) to (F)]. (A) Square, (B) angled (all angles other than 90 degrees), (C) shoe, (D) twin radius—equal (all angles), (E) twin radius—unequal (all angles), and (F) twin—square.

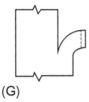

Fig. 14.8 Branch for rectangular ducts, type (G). (G) Branch (for all turning angles; splitters in accordance with Table 13.7).

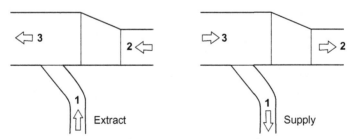

Fig. 14.9 Typical arrangement of supply and extract branches.

Loss coefficients K_{31} for twin radius tee pieces with equal area ducts are given in Table 14.15. For *rectangular* ducts the loss coefficients should be increased by 10% over the circular duct loss.

Table 14.13 Supply branch (conical or shoe) correction factors C_{31} for square branches

Duct area ratio	Flow rate ratio, $\dot{V}_1/\dot{V}_3$						
	0.2	0.3	0.4	0.5	0.6	0.7	0.8
0.4	0.95	0.86	0.80	0.75	0.67	0.64	0.60
0.6	0.97	0.92	0.87	0.82	0.77	0.72	0.68
0.8	0.98	0.95	0.90	0.84	0.81	0.77	0.73

Note: factors are applied to the loss coefficients K_{31} in Fig. 14.11, using Eq. (14.8).

Table 14.14 Loss coefficients K_{32} for supply branch, flow to main

		Flow rate ratio, $\dot{V}_2/\dot{V}_3$					
		0.2	0.3	0.4	0.5	0.6	0.7
	K_{32}	0.22	0.15	0.09	0.03	–	–

Note: coefficients are independent of type of off-take at branch 1; duct area ratio $A_2/A_3 = 1$.

Table 14.15 Loss coefficients K_{31} for supply 90 degrees twin radius tee, equal area ducts

		Flow rate ratio, $\dot{V}_1/\dot{V}_3$						
		0.2	0.3	0.4	0.5	0.6	0.7	0.8
	K_{31}	0.48	0.41	0.35	0.30	0.27	0.23	0.20

Example 14.9

Determine the pressure losses for a 90 degrees branch shown in Fig. 14.10 at a standard air density of 1.2 kg/m³:

(1) for a 90 degrees branch with sharp edges
(2) for a 45 degrees branch (shoe)

Solution

The relevant data is set out in the following table:

Duct	Air flow rate, $\dot{V}$ (m³/s)	Duct diameter, d (m)	Duct area, A (m²)	Velocity, $\bar{v}$ (m/s)	Velocity pressure, p_v (Pa)
1	0.1	0.16	0.02	–	–
2	0.3	0.25	0.049	–	–
3	0.4	0.25	0.049	8.15	39.8

Loss to branch 1

(1) The area ratio $A_1/A_2 = 0.02/0.049 = 0.41$
The flow rate ratio $\dot{V}_1/\dot{V}_2 = 0.1/0.4 = 0.25$
From Fig. 14.11A the loss coefficient $K_{31} = 1.0$
∴ the pressure loss $= K_{31}p_{v3} = 1.0 \times 39.8 = 39.8\,\text{Pa}$

(2) From Fig. 14.11B $K_{31} = 0.52$
∴ the pressure loss $= K_{31}p_{v3} = 0.52 \times 39.8 = 20.7\,\text{Pa}$

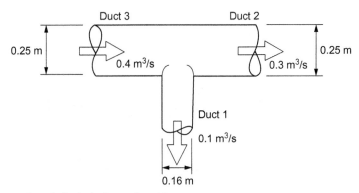

Fig. 14.10 Supply branch: Example 14.9.

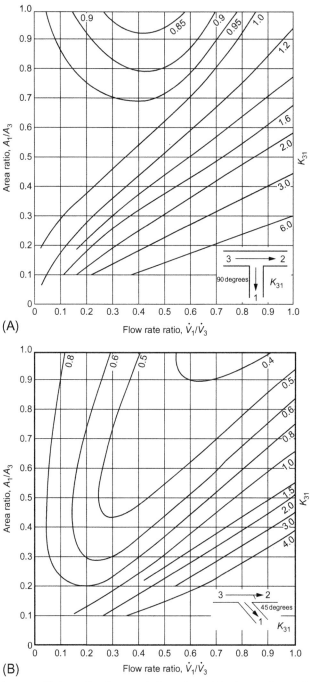

Fig. 14.11 Loss coefficients K_{31} for supply branches with sharp edges (A) 90 degrees branch, (B) 45 degrees branch. *(Reproduced from D.S. Miller, Internal Flow Systems, BHRA Cranfield, with permission.)*

For *extract* ducts the loss coefficients K_{13} and K_{23} refer to the velocity pressure in the duct with the *combined flow*; this duct is designated as duct 3.

As with the supply duct branch pieces, the loss coefficient refers to the velocity pressure in the duct with the combined flow:

K_{13} refers to the loss from branch 1 to duct 3;

K_{23} refers to the loss from branch 2 to duct 3.

The loss coefficients K_{13} and K_{23} for 45 and 90 degrees circular duct branch connections are given in Figs 14.12 and 14.13. These diagrams give the contours of constant loss coefficients, and they are typical of the data given by Miller. Loss coefficients for a combining tee are given in Table 14.16.

Due to the suction effect caused by the flow in the main, the loss factor can take a negative value. This means that when the total pressure drop is calculated for the branch duct, the loss at the branch is *subtracted* from the total of the other losses in the branch.

- For conical or shoe connections, the loss coefficients are *reduced* by 10%.
- For rectangular ducts the loss coefficients should be *increased* by 10%.

Example 14.10

Determine the pressure losses for a 45 degrees extract suction branch piece shown in Fig. 14.14 with a standard air density of 1.2 kg/m^3 using the following data:

Duct	Air flow rate, $\dot{V}$ (m³/s)	Duct area, A (m²)	Velocity, $\bar{v}$ (m/s)	Velocity pressure, p_v (Pa)
1	0.1	0.02	5.0	—
2	0.3	0.049	10.0	—
3	0.4	0.049	12.5	91.8

Solution

Flow rate ratio: $\dot{V}_1/\dot{V}_2 = 0.1/0.5 = 0.2$

Area ratio: $A_1/A_2 = 0.2/0.4 = 0.5$

From Fig. 14.12 the loss coefficient for the suction branch to the main is K_{13} is minus 0.2.

Therefore the pressure *gain* across branch is obtained:

$$\Delta p = K_{13}p_{v3} = 0.2 \times 94 = 18.8\,\text{Pa}$$

From Fig. 14.12B, the *loss* coefficient in main K_{23} is 0.2.

Therefore the pressure loss in the main duct is:

$$\Delta p = K_{23}p_{v3} = 0.2 \times 94 = 18.8\,\text{Pa}$$

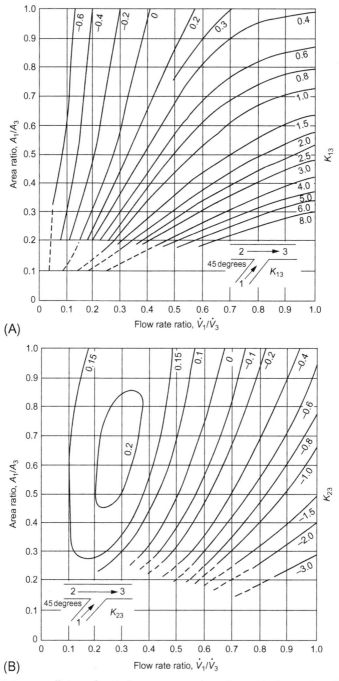

Fig. 14.12 Loss coefficients for 45 degrees extract branches with sharp edges (A) mains loss K_{13} (B) branch loss K_{23}. *(Reproduced from D.S. Miller, Internal Flow Systems, BHRA Cranfield, with permission.)*

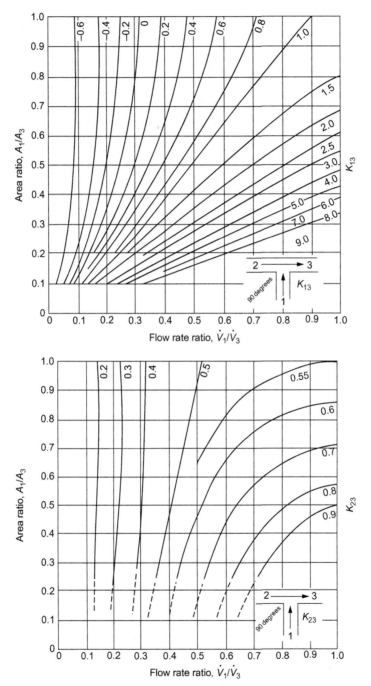

Fig. 14.13 Loss coefficients for 90 degrees extract branches with sharp edges (A) mains loss K_{13} (B) branch loss K_{23}. *(Reproduced from D.S. Miller, Internal Flow Systems, BHRA Cranfield, with permission.)*

Table 14.16 Loss coefficients K_{13} for extract 90 degrees twin radius tee, with equal area ducts

		Flow rate ratio, $\dot{V}_1/\dot{V}_3$						
		0.2	0.3	0.4	0.5	0.6	0.7	0.8
K_{13}	–		0.05	0.12	0.18	0.21	0.23	0.27

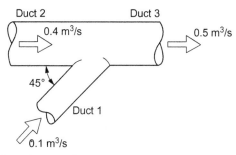

Fig. 14.14 Extract branch: Example 14.10.

Duct Entries

Pressure loss coefficients for duct entries and intake louvres are given in Tables 14.17 and 14.18.

Example 14.11

Determine the pressure loss across the intake louvre with dimensions 400 mm and 300 mm. The air flow rate is 0.4 m³/s at standard air density.

Solution

The mean velocity is:

$$\bar{v} = \frac{\dot{V}}{A} = \frac{0.4}{0.4 \times 0.3} = 3.3\,\text{m/s}$$

The velocity pressure is:

$$p_v = 0.6\bar{v}^2 = 5.4 \times 3.3^2 = 6.5\,\text{Pa}$$

From Table 14.18, and assuming a type c louvre with a free area ratio of 0.6, the loss coefficient $= 1.8$. Therefore the pressure loss across intake louvre is:

$$\Delta p = K_i p_v = 1.8 \times 6.5 = 11.8\,\text{Pa}$$

Table 14.17 Loss coefficients K_i for circular duct inlets

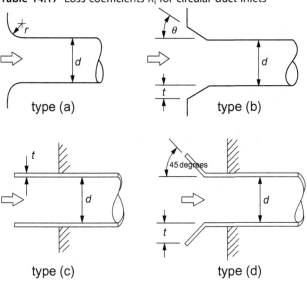

t/D or r/D	(a)	(b) θ = 30 degrees	(b) θ = 45 degrees	(c)	(d)
0.00	0.50	0.50	0.50	1.00	1.00
0.05	0.20	0.25	0.30	0.58	0.58
0.10	0.08	0.17	0.23	0.54	0.40
0.20	0.06	0.12	0.20	0.54	0.22
0.30	0.06	0.12	0.20	0.54	0.22

Note: for rectangular ducts, increase corresponding circular duct coefficient by 10%.

Table 14.18 Loss coefficients for 45 degrees intake louvres

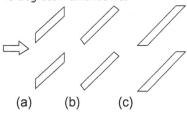

Free area ratio	Type of louvre (a)	(b)	(c)
0.5	6.0	4.5	3.0
0.6	3.6	3.0	1.8
0.7	2.4	2.1	1.1
0.8	1.6	1.4	0.7

Duct Discharges

Pressure loss coefficients for duct discharges are given in Table 14.19. These are losses for discharges into rooms, plenums, and large open spaces and are in addition to any grille covering the outlet. The air discharges to datum pressure which is most often at atmospheric pressure.

Example 14.12

Determine the pressure loss at a circular, plain 300 mm diameter supply discharge to a room. The flow rate is 0.5 m³/s at standard air density.

Solution

The mean velocity is:

$$\bar{v} = \frac{\dot{V}}{A} = \frac{0.5}{\pi 0.3^2/4} = 7.07\,\text{m/s}$$

The velocity pressure is:

$$p_v = 0.6\bar{v}^2 = 0.6 \times 7.07^2 = 30\,\text{Pa}$$

From Table 14.19, the pressure loss factor is 1.0.
Therefore the pressure loss at the outlet is:

$$\Delta p = K_{ex}p_v = 1.0 \times 30 = 30\,\text{Pa}$$

Table 14.19 Loss coefficient K_{ex} for discharge to a plenum or to atmosphere

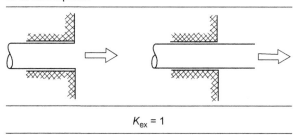

$$K_{ex} = 1$$

DUCT SIZING PROCEDURES

Reference to the friction chart, and as indicated in Example 14.1, suggests the two most commonly used methods of obtaining the size of a duct to carry a given flow rate, i.e., either by setting a duct velocity or by

Table 14.20 Recommended air velocities and pressure drops for sizing low velocity ductwork systems
(a) Air velocities in m/s

Item	Type of building		
	Residential hospitals hotels	Commercial schools theatres	Industrial
Outdoor air intake	2.0–3.0	2.0–3.0	2.0–3.0
Fan inlet	3.0–4.0	3.5–4.5	4.5–5.5
Fan outlet	5.0–8.0	6.5–10.0	8.0–12.0
Main duct	3.5–4.5	5.0–6.5	6.0–9.0
Branch duct	2.5–3.5	3.0–4.5	4.0–5.0
Outlet duct	2.0–2.5	3.0–3.5	3.5–4.5

(b) Pressure drops per unit length of duct

Application	Pressure drop per unit length of duct (Pa/m)
Concert halls, broadcasting studios, recording studios	0.4
Hospitals, libraries, theatres	0.65
General offices, department stores	0.8
Factories and workshops	1.0–1.5

setting a pressure drop per unit length. These two methods are usually referred to as:

- sizing by velocity;
- sizing by pressure drop per unit length.

The duct sizing procedure using the second of these methods is demonstrated by the following example. (The sizing procedure would be similar for the second method.) The example also illustrates a number of important points which are discussed in the solution. Selection of suitable velocities and pressure drops for low velocity systems are given in Table 14.20.

Example 14.13

Determine the total system pressure loss for system of rectangular ductwork shown in Fig. 14.15. Each terminal unit supplies 0.5 m/s of air to the air conditioned spaces. The ductwork is to be sized on a pressure drop of 1.0 Pa/m length of duct with loss coefficients for fittings obtained from the relevant tables.

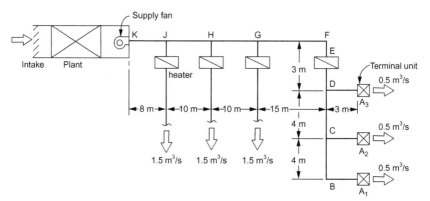

Fig. 14.15 Ductwork layout for Example 14.12.

Plant pressure drops:

terminal units   20 Pa
heaters   45 Pa
plant (to fan suction) 250 Pa

Solution

It is often convenient to calculate the losses in a supply system *against the flow direction*, in order to generate subtotals. This enables the index run to be identified, at the same time providing information for balancing the branches ducts. (With an extract system the sizing procedure will be in the same direction as the flow.) The majority of the calculations are completed in Table 14.21 and relevant notes are as follows:

- The pressure losses of the terminal unit, heater battery, and central plant are obtained from manufacturer's information.
- Where the final discharge from the system is in the form of a jet, the velocity pressure from the discharge outlet is a system pressure loss and should be included in the sum of the total losses with which the fan has to deal. In systems supplying air for comfort air conditioning this will be a relatively small loss, as the outlet velocity will be about 2 m/s and the velocity pressure if omitted from the total system pressure loss would not affect the system/fan performance. However it is good practise to include it, so that when it is relatively large compared with the total system loss (for example in an industrial exhaust system with high discharge velocity) it may not be omitted.
- Rectangular duct sizes have been used such that the equivalent circular duct diameter gives a friction pressure loss of approximately 1.0 Pa/m of duct length. Duct lengths are measured along the axis of the duct; the

Table 14.21 Duct sizing calculations for Example 14.13

Item Fitting duct plant	Flow rate $\dot{V}$	Duct size Rectangular, $b \times W$ (m×m)	Circular (equiv) D (m)	Velocity $\bar{V}$ (m/s)	Fitting Velocity pressure, p_v	Loss coeff., K	Straight duct Length l (m)	Pressure loss/m Δp_f (Pa)	Item $K p_v$ or $l \times p_v$	Pressure loss (Pa) Subtotal	Accumulated total
Terminal A₁	0.5				—	—			20		
Duct A₁BC	0.5	0.4 × 0.25	0.346	5.0	—	—	7.0	1.0	7		
Bend B	0.5			5.0	15.0	0.42	—	—	7		
Branch C	1.0/0.5			6.25	21.4	0.08	—	—	2	34	
Duct CD	0.5	0.4 × 0.4	0.441	6.25	—	—	4.0	1.1	4		
Branch D	1.5/1.0			6.25	21.4	0.05	—	—	1	5	39
Duct DEFG	1.0	0.4 × 0.6	0.537	6.25	—	—	18.0	0.8	14		
Heater E	1.5			6.25	—	—	—	—	45		
Bend F	1.5			6.25	21.4	0.24	—	—	6		
Branch G	3.0/1.5			8.30	41.3	0.05	—	—	2	67	106
Duct GH	3.0	0.6 × 0.6	0.661	8.3	—	—	10.0	1.2	12		
Branch H	4.5/3.0			9.4	53.0	0.05	—	—	3	15	121
Duct HJ	4.5	0.8 × 0.6	0.848	9.4	—	—	8.0	1.0	11		
Branch J	6.0/4.5			9.4	53.0	0.05	—	—	3	14	135
Duct JK	6.0	0.8 × 0.8	0.881	9.4	—	—	—	—	8		
Fan discharge	6.0			a	a	a	—	—	a	8	143
Plant	6.0				—	—	—	—	250		
Intake grille	6.0	2.0 × 1.2	—	2.5	3.73	3.0	—	—	9	259	402
Terminal A₂	0.5			—	—	—	—	—	20		
Duct A₂C	0.5	0.4 × 0.25	0.346	5.0	—	—	3.0	0.9	3		
Branch C	1.0/0.5			6.25	23.4	0.2	—	—	5	28	

[a] Pressure loss not available until the fan has been selected and the discharge designed (see text).

fittings are assumed to have zero length (in accordance with the definition of loss coefficients).

The following points should be noted:

- The pressure losses are given as integers.
- The total pressure drop in the system is 402 Pa. As the fan has not yet been sized, this total does not include the loss in the fan discharge expansion. The fan duty is therefore selected for a duty of 6 m³/s at 402 Pa on fan static pressure, as described in Chapter 15.
- Reynolds number and interaction effects between fittings have not been included in these calculations.

Index Run and System Balance

Referring to Example 14.13 and Table 14.21, from terminal A_2 the duct A_2C has been sized for the same friction pressure drop as the duct supplying terminal A_2. The pressure loss up to the point of off-take has a sub-total of 28 Pa, which is 6 Pa less than duct A_1BC and the two branch ducts are *out of balance*. The index run is therefore taken as A_1 through to the fan discharge. The out-of-balance pressure drop in branch CA_1 (and similarly the other branches) can be dealt with in one of three ways:

- resize the duct at a higher velocity so that the duct and fittings together absorb pressure drop of 34 Pa;
- include a fitting such as a perforated plate, sized to give a net pressure loss of 6 Pa; include a damper which can be adjusted on site, to provide the out-of-balance pressure drop.
- Reynolds number and interaction effects between fittings have not been included in the calculations.

Though a balance at the design stage is theoretically possible with either of the first two methods, the calculations are based on published pressure loss coefficients. It is most unlikely that this can be achieved because the duct friction loss is based on an assumed roughness coefficient and the loss coefficients are experimental values obtained, for the most part with fully developed, turbulent flow conditions under laboratory conditions. The difference between these and the actual losses can be significant, so that in practise a balance cannot be achieved by design calculations. The third solution of placing dampers in the network should therefore be adopted; their location has to be based on the principles of proportional balancing procedures described in Chapter 16.

LAYOUT CONSIDERATIONS

Ductwork systems have to be integrated into the building design and coordinated with other services. There are a number of ways in which they can be arranged; often the simplest method is to accommodate them above a false ceiling in a central corridor, with connections at high level facing the external windows. In this case, access to services is relatively easy. Other types of distribution include the following, individually or in combination:

- vertical risers, e.g., toilet extract;
- horizontal perimeter ducts to serve window sill units; the duct runs may be under the window or in a bulkhead at ceiling level on the floor below;
- ducts above false ceilings, supplying ceiling diffusers; the space above the ceiling can be used for plenum extract, a common approach when extract air lighting fittings are installed;
- air handling units above the false ceiling;
- ducts under a false floor;
- inter-floor plant space.

Balancing Requirements

The ductwork distribution systems should be designed in such a way as to assist the balancing procedures and included facilities for measuring and regulating the air flow rates. These requirements are described in Chapter 16.

Connections to Plant Items

As with the fan suction and discharge connections (see Chapter 15), duct connections to plant items which produce asymmetrical velocity profiles at the upstream face of the equipment may lead to increased pressure drop, increased noise generation, lower efficiency and moisture carry-over. They should therefore, be designed to ensure an even flow over such items as filters, heating and cooling coils, humidifiers, dampers, and acoustic attenuators.

The fresh air intake grille should be positioned away from the exhaust outlets to minimize the risk of bringing in humid and/or contaminated air. To avoid heat gains from the plant room itself, the intake should be connected, either by sheet metal or by builders' work, to the first item of plant in the system.

Smoke and Fire Precautions

The ductwork system, including any linings, must be designed and installed to minimize the risk of smoke and fire spreading through the building via the ductwork itself. The materials used should be noncombustible or, at the very least, difficult to ignite, possess a good rating against the surface spread of flame and not generate smoke or toxic fumes in the presence of fire or heat. Special attention should be given to the insulating materials and adhesives, flexible connections, filters, and acoustic linings.

When a duct passes through a fire resisting builders work enclosure, any holes through which the duct passes should be as small as possible and when installed sealed with fire-stopping material. Similar treatment should be applied to ducts which pass through floors and partitions designed to prevent the spread of fire and smoke.

To reduce the risk of fire and smoke spreading through the ductwork, fire dampers must be installed where the ductwork passes through a fire partition. Detectors for smoke and fire warning signals are often included in the ductwork. It is important to follow local authority and fire brigade regulations in all these aspects of design.

SYMBOLS

A duct or pipe cross section area
B duct breadth
C_{b-b} bend interaction effect factor
C_e loss factor for expansions exhaust discharge
C_{Re} Reynolds number correction factor
C_{31} branch correction factor
C_r curve ratio for guide vanes in rectangular ducts
D diameter
D_e equivalent diameter of rectangular duct
K pressure loss coefficient
K_b pressure loss coefficient, bend
K_c pressure loss coefficient, contraction
K_e pressure loss coefficient, expansion
K_{es} pressure loss coefficient, expansion with short outlet pipe
K_{ex} pressure loss coefficient, exhaust discharge
K_i pressure loss coefficient, correction factor
K_{31} pressure loss coefficient, supply branch
K_{32} pressure loss coefficients, supply branch
K_{13} pressure loss coefficient, extract branch
K_{23} pressure loss coefficients, extract branch
L duct length
L_s spacer length (L/D)
N length of expansion

n	number of vanes in mitred
p_v	velocity pressure
Re	Reynolds number
r	radius
t	air temperature
$\dot{V}$	volume flow rate
$\bar{v}$	mean duct or pipe velocity
W	duct width
Δp	pressure drop
Δp_f	pressure loss due to friction
ρ	air density
α	angle of expansion vanes, from axis
θ	angle of an expansion

SUBSCRIPTS

1, 2   relates to specific duct or pipe section

ABBREVIATIONS

AR   aspect ratio of rectangular duct (b/W)

REFERENCES

[1] CIBSE Guide C, Reference Data, 2007.
[2] BS EN 1506:2007 Ventilation for buildings. Sheet metal air ducts and fitting with circular cross-section; dimensions.
[3] CIBSE Guide C, op. cit.
[4] D.S. Miller, Internal Flow Systems, BHRA Fluid Engineering (originally published 1978, reprinted 2008).
[5] ASHRAE Handbook, Fundamentals, 2013 (Chapter 21).
[6] HESA, Specification for Sheet Metal Ductwork: Low, Medium and High Pressure/Velocity Air Systems, 2014.
[7] D.S. Miller, op.cit.
[8] Ibid.

CHAPTER 15

Fans

Fan selection is necessarily important to the overall performance of air conditioning systems. The required duty is obtained by an analysis of the flow rates and pressure losses in the ductwork system as described in the previous chapter. The choice of an appropriate unit will then depend on the application, together with considerations such as plant arrangement, space available for the installation, and corrosive and hazardous gases. Since the annual energy cost for fan operation represents a relatively high proportion of the total energy costs of an air conditioned building, it is also important to consider the fan's overall efficiency.

CHARACTERISTICS

By running a fan at constant speed and measuring the pressure developed and the power input at different flow rates, the manufacturer obtains data on the fan's performance. These measurements, together with the computed efficiency, when graphed or tabulated, are known as the *fan's characteristics*. The tests are carried out in accordance with recognized national-international standards [1].

FAN TYPES

The following are the five main categories of fans used for a variety of applications in air conditioning and ventilation systems:
- Centrifugal,
- Axial,
- Mixed flow,
- Propeller,
- Cross flow.

These fans are described below, together with graphs of their typical characteristic curves of:
- pressure versus flow rate,
- efficiency versus flow rate,
- power versus flow rate.

Not every part of a fan's characteristic will give satisfactory operation in the system. Therefore, on each of the fan curves, an indication is given of the *normal working range*.

Centrifugal Fans

Centrifugal fans may be single or double inlet and are either belt-driven or direct-driven by an electric motor. A single-inlet fan is illustrated in Fig. 15.1.

Centrifugal fans are classified by their impeller blade shape, i.e.,

- backward curved,
- radial,
- forward curved.

The blade shapes and characteristic curves for these fan types are illustrated in Figs 15.2–15.4.

The *backward curved* impeller, with 6–16 blades, has a power characteristic that rises to a maximum at the middle of its flow rate range and then falls at its highest flow rates. This is known as a nonoverloading power characteristic. These fans have efficiencies of up to 90% where aerofoil section blades are used. They are most commonly used when continuous high efficiency operation is required.

The *radial-* or *paddle-bladed* impeller, with 6–16 blades, has a continuously rising or overloading power characteristic. With flat blades, this fan type has an efficiency up to 60%, with blades slightly curved at the heel, up to 75%. It is most often used where corrosion and blade wear may be a problem, as in industrial exhaust systems, the blades being self-cleaning and often replaceable.

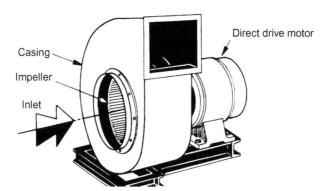

Fig. 15.1 Single-inlet centrifugal fan.

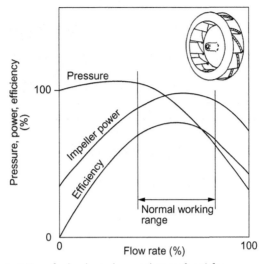

Fig. 15.2 Characteristics of a backward curved centrifugal fan.

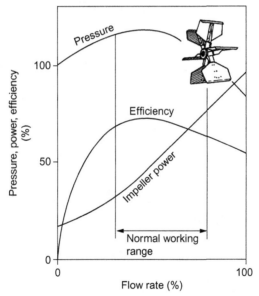

Fig. 15.3 Characteristics of a radial centrifugal fan.

The *forward-curved* fan has a 40–60 bladed impeller with an overloading power characteristic and an efficiency of up to 75%. It is compact in size, and because of this, it is the most commonly used centrifugal fan for AC and mechanical ventilation systems.

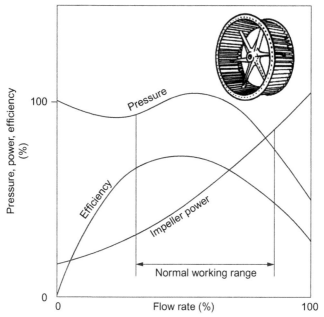

Fig. 15.4 Characteristics of a forward-curved centrifugal fan.

The overall design of centrifugal fans makes a single-inlet fan a suitable choice when the plant installation requires a right-angled turn. Double-inlet fans are usually installed in plenum chambers and are often used in this way in packaged AC units.

Axial Flow Fans

An axial flow fan has an impeller with 6–12 aerofoil section blades with a non-overloading power characteristic with efficiencies up to 85%. To achieve these high efficiencies, guide vanes are used either to give prerotation on the upstream side or to remove air rotation from the downstream side of the fan. An alternative method of improving efficiency is to have two stages rotating in opposite directions (contrarotating). The second impeller then acts in a similar manner to the upstream guide vanes. The fan is most often directly driven by a synchronous speed motor inside the fan casing though the fan can also be belt-driven by an external motor where the installation requires it, e.g., in hot, dirty, or corrosive airstreams (see Fig. 15.5).

Variable-pitch axial flow fans are used for VAV air conditioning systems. The pitch angle of the blades are varied through a controller, e.g., a pneumatic actuator, with increasing blade angle for increasing flow rates and vice versa for reducing flow rates. The characteristics for a typical fan are given in Fig. 15.6.

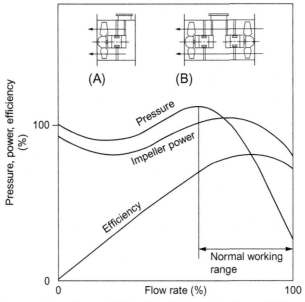

Fig. 15.5 Characteristics of an axial flow fan. (A) Single stage and (B) two-stage with contrarotating impellers.

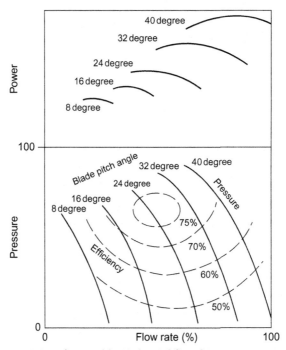

Fig. 15.6 Characteristics of a variable-pitch axial flow fan.

Axial fans are compact and, relatively easily, incorporated in the system. It can be placed at low or high level in the plant room, in false ceilings, and in either vertical or horizontal ducts. The fan's configuration makes it suitable for installations in which the intake and discharge connections are opposite to one another.

Mixed Flow Fans

Mixed flow fans are intermediate between the centrifugal and axial fan types in pressure development and compactness. The air path through the impeller has a radial and an axial component. These fans are normally constructed with the intake and discharge on the same axis, though they can be fitted with a radial discharge. Efficiencies are up to 80% with a nonoverloading characteristic.

Propeller Fans

The propeller fan is a comparatively simple form of a fan with a sheet metal, 3–6 bladed impeller directly driven by a motor mounted in the airstream. Efficiencies up to 75% can be achieved, though pressure development is low and overloading can occur if the fan is installed in a system with too high a resistance. Performance can be improved by using aerofoil blades that give efficiencies up to 80%. The fan is useful for low-pressure drop systems, e.g., *through-the-wall* kitchen extract. It is also used extensively for unit equipment such as refrigeration air-cooled condensers and for cooling towers (Figs 15.7 and 15.8).

Cross Flow Fans

The cross flow fan has a multibladed, cylinder-like impeller with blade shape similar to the forward-curved centrifugal fan. The ends of the impeller are blanked off and, because of the casing shape, air enters along one of the cylindrical surfaces and discharges from the other. The maximum efficiency is about 40%. The principal application is in small domestic electric heaters (Fig. 15.9).

FAN LAWS

It is not practical to test every fan, and the performance of geometrically similar fans is obtained by applying the fan laws. These laws apply to a particular point on a fan characteristic and cannot be used to predict other

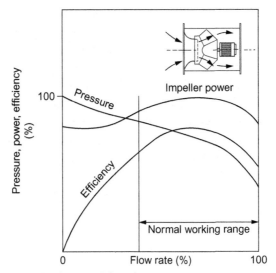

Fig. 15.7 Characteristics of a mixed flow fan.

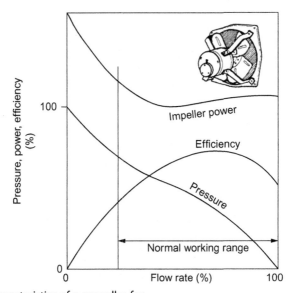

Fig. 15.8 Characteristics of a propeller fan.

points. The fan volume flow rate and pressure are related to the fan speed, diameter, and air density according to the following equations:

Fan flow rate is given by:

$$\dot{V} = C_{FV} n D^3 \tag{15.1}$$

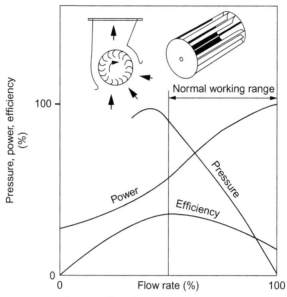

Fig. 15.9 Characteristics of a cross flow fan.

Fan pressure is given by:

$$p_F = C_{Fp} n^2 D^2 \rho \qquad (15.2)$$

where C_{FV} is the constant of proportionality for fan flow rate
C_{Fp} is the constant of proportionality for fan pressure.
Airpower is given by:

$$P_a = \dot{V} p_F \qquad (15.3)$$

Combining Eqs (15.1), (15.2),

$$P_a = C_{Fp} n^3 D^3 \rho \qquad (15.4)$$

where C_{Fp} is the constant of proportionality for fan power
Fan efficiency η is given by:

$$\eta = \frac{P_a}{P_i} \qquad (15.5)$$

where P_i is the impeller power
Notes:
1) The constants of proportionality C_{Fv}, C_{Fp}, and C_{FP} are specific for the type of fan and the point of operation.
2) The *overall* fan efficiency must be related to the motor power input, which takes account of motor and drive efficiencies.

Example 15.1

A centrifugal fan with a 0.6 m diameter impeller, running at 13 rev/s, delivers 2 m³/s of air with an air density of 1.2 kg/m³, at a total pressure of 500 Pa. Determine the duty of a geometrically similar fan with a 0.8 m diameter impeller running at 20 rev/s if the air density is the same for both fans. Determine the airpower of the second fan.

Solution

To obtain constant of proportionality for flow rate, use Eq. (15.1):

$$\dot{V} = C_{FV}\, n\, D^3$$

$$2 = C_{FV} \times 13 \times 0.6^3$$

$$\therefore C_{FV} = 0.712$$

The new fan flow rate is given by Eq. (15.1):

$$\dot{V} = 0.712 \times 20 \times 0.8^3 = 7.29\, m^3/s$$

To obtain the constant of proportionality for pressure, Eq. (15.2) gives:

$$p_F = C_{Fp}\, n^2 D^2 \rho$$

$$500 = C_{Fp} \times 13^2 \times 0.6^2 \times 1.2$$

$$\therefore C_{Fp} = 6.85$$

From Eq. (15.2), the new fan pressure is:

$$p_F = 0.85 \times 20^2 \times 0.8^2 \times 1.2 = 2104\, Pa$$

Airpower is given by Eq. (15.3):

$$P_a = \dot{V} p_F = \frac{7.29 \times 2104}{1000} = 15.3\, kW$$

Example 15.2

For the centrifugal fan in Example 15.1, with an increase in impeller size to 0.8 m, determine the new duty and airpower if the air density reduces to 1.05 kg/m³:

Solution

The air volume flow rate is unaffected by air density:
∴ the flow rate is 7.29 m³/s.
From Example 15.1, the constant of proportionality for pressure C_{Fp} is 6.85.

The new fan pressure is given by Eq. (15.2):

$$p_F = C_{Fp}n^2D^2\rho$$
$$= 6.85 \times 20^2 \times 0.8^2 \times 1.05 = 1841\,\text{Pa}$$

New airpower is given by Eq. (15.3):

$$P_a = \dot{V}p_F = \frac{7.29 \times 1841}{1000} = 13.4\,\text{kW}$$

Operating Point

In an airflow system consisting of ducts, fittings, and plant items, there will be a system total pressure loss Δp_t versus the total flow rate $\dot{V}$. This determines the system characteristic that is assumed to obey the square law (Eq. 13.16):

$$\Delta p_t = r_t\,\dot{V}^2 \tag{15.6}$$

where r_t is known as the *system total resistance*.

A fan is selected for the design duty, and the system and fan characteristics intersect, as shown in Fig. 15.10, to give the fan and system operating point. The volume flow rate and pressure at the operating point may be obtained either graphically or by solving simultaneously the system and fan characteristic equations.

For most fans, the fan volume/pressure characteristic in the normal working range can be defined by a quadratic equation of the form:

$$p_F = A' - B'\,\dot{V}^2 \tag{15.7}$$

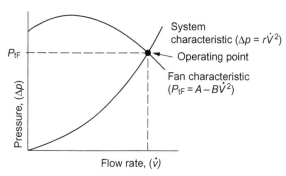

Fig. 15.10 Operating point of a fan.

A' and B' are the fan characteristic constants determined from a curve fit of the manufacturer's data, and their values apply to a particular fan at a given speed and air density.

Example 15.3

Determine the operating point for a fan that is connected to a system with a design duty of $2\,m^3/s$ and a total pressure drop of 440 Pa. The fan characteristic constants for Eq. (15.7) are $A' = 600$ and $B' = 20$.

Solution

The system resistance is obtained using Eq. (15.6):

$$\Delta p = r_t \, \dot{V}^2$$

$$440 = r_t \, 2^2$$

$$\therefore r_t = 440/4 = 110$$

The system characteristic is given by:

$$\Delta p = 110 \, \dot{V}^2$$

The fan characteristic is given by Eq. (15.7):

$$p_F = A' - B' \, \dot{V}^2$$

At the operating point:

$$\Delta p = p_F$$

$$\therefore 110 \, \dot{V}^2 = 600 - 20 \, \dot{V}^2$$

$$\therefore \dot{V} = \sqrt{\frac{600}{110 + 20}} = 2.15\,m^3/s$$

The fan pressure is then given by:

$$p_F = 600 - 20 \, \dot{V}^2 = 600 - 20 \times 2.15^2 = 508\,Pa$$

The operating point of the fan/system is then $2.15\,m^3/s$ at a pressure of 508 Pa.

ADJUSTMENT OF TOTAL FLOW RATE

If the airflow rate is not at the design level when the system has been balanced to the design flow rate, adjustment of the total flow rate in constant flow systems may be achieved by one of the following methods:

For reducing flow rates,

- throttling damper,
- inlet guide vanes (centrifugal fans).

For increasing or reducing flow rates,

- change of fan speed,
- change of pitch angle (axial fans).

Throttling Damper

The simplest method of reducing the total flow rate is to restrict the flow with a throttling damper. This has the effect of increasing the system resistance, the operating point *retreating* along the fan characteristic curve. This is illustrated in Fig. 15.11.

Example 15.4

Determine the pressure drop across a main throttling damper to reduce the flow rate in Example 15.3 to the design duty.

Solution

Using the fan characteristic equation, at the design flow rate of $2 \, \mathrm{m^3/s}$, the fan pressure is given by:

$$p_F = 600 - 20 \times 2^2 = 520 \, \mathrm{Pa}$$

The system pressure drop at the design flow rate $= 440$ Pa.
The pressure drop across the damper $= 520 - 440 = 80$ Pa.

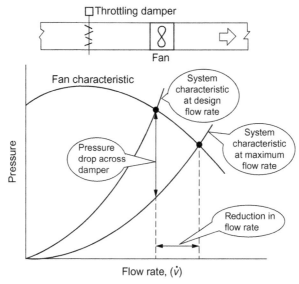

Fig. 15.11 Reduction in flow rate with a throttling damper.

Speed Control

Since fan power is proportional to the *cube* of fan speed, reducing the speed of the fan is the most energy-efficient method of reducing the flow rate. For a belt-driven centrifugal fan, this may be achieved either by changing the pulley size or by incorporating continuously variable pulleys in the fan drive.

Example 15.5

Determine the new fan speed to reduce the flow rate in Example 15.4 to the design duty if the original speed was 13 rev/s. Compare the airpower requirement with that using a throttling damper in Example 15.3.

Solution

From Example 15.3, the airflow rate is 2.15 m^3/s.
The design duty is 2 m^3/s.
Referring to Eq. (15.1), the flow rate is proportional to speed, and the reduced speed is given by:

$$n = \frac{2}{2.15} \times 13 = 12.1 \,\text{rev/s}$$

Referring to Eq. (15.4), the airpower is proportional to (speed)^3. If the fan is now made to run at 12.1 rev/s, the percentage reduction in fan *air* power is given by:

$$\left[1 - \left(\frac{12.1}{13} \right)^3 \right] \times 100 = 19.6\%$$

SERIES AND PARALLEL OPERATION

Series Operation

Two fans, of equal duty, which are connected in series, have a combined flow/pressure characteristic that doubles the pressure at the same flow rate compared with a single fan. This arrangement is most suitable for high-resistance systems; the combined characteristic is illustrated in Fig. 15.12.

Parallel Operation

Two fans of equal duty, which are working in parallel, have a combined flow/pressure characteristic that doubles the flow rate at the same pressure, compared with a single fan. This arrangement is most suitable for low-resistance systems; the combined characteristic is illustrated in Fig. 15.13.

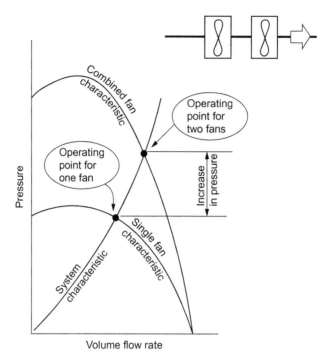

Fig. 15.12 Characteristic of two fans in series.

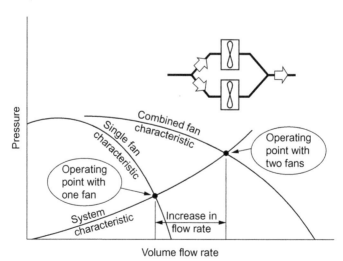

Fig. 15.13 Characteristic of two fans in parallel.

However, these fan arrangements will not produce in the system *double the pressure* or *double the flow rate* as, with a single fan, the operating point for the combined fan characteristics will depend on the characteristic of the system in which the fans are installed.

Example 15.6

Two identical fans, with characteristic constants $A' = 500$ and $B' = 10$, are connected in series to a system with a resistance of 120. Determine the combined duty of the two fans.

Solution

The single fan characteristic is given by Eq. (15.7):

$$p_F = A' - B'\,\dot{V}^2$$
$$= 500 - 10\,\dot{V}^2$$

Therefore, the characteristic of the two fans in series is given by:

$$p_F = 2\left(500 - 10\,\dot{V}^2\right)$$

The system characteristic is given by:

$$\Delta p = 120\,\dot{V}^2$$

At the operating point:

$$\Delta p = p_F$$
$$\therefore 120\,\dot{V}^2 = 2\left(500 - 10\,\dot{V}^2\right)$$

$$\therefore \dot{V} = \sqrt{\frac{1000}{(120 + 20)}} = 2.67\,\mathrm{m^3/s}$$

The operating pressure is therefore given by:

$$\Delta p = 120\,\dot{V}^2 = 120 \times 2.67^2 = 856\,\mathrm{Pa}$$

The series fans' duty is 2.67 m³/s at a pressure of 856 Pa; this compares with the single fan operating at a pressure of 461 Pa.

Example 15.7

Two identical fans, with characteristic constants $A' = 600$ and $B' = 20$, are connected in parallel to a system with a resistance of 12. Determine the combined duty of the fans in series.

Solution

The single fan characteristic is given by Eq. (15.7):

$$p_F = 500 - 10\,\dot{V}^2$$

$$\therefore \dot{V} = \sqrt{\frac{500 - p_F}{10}}$$

The characteristic of two fans in parallel is given by:

$$\dot{V} = 2 \times \sqrt{\frac{500 - p_F}{10}}$$

From the system characteristic, the flow rate is also given by:

$$\dot{V} = \sqrt{\frac{\Delta p}{12}}$$

At the operating point, $\Delta p = p_F = p$

$$\therefore 2 \times \sqrt{\frac{500 - p}{10}} = \sqrt{\frac{p}{12}}$$

$$\therefore \Delta p = 414\,\text{Pa}$$

$$\therefore \dot{V} = \sqrt{\frac{414}{12}} = 5.87\,\text{m}^3/\text{s}$$

The parallel fans' flow rate is 5.87 m^3/s at a pressure of 414 Pa; this compares with the single fan flow rate of 4.77 m^3/s.

CONTROL FOR A VAV SYSTEM: CENTRIFUGAL FANS

Inlet Guide Vanes

This method has been a popular form of control where centrifugal fans are used. The guide vane assembly may be either bolted onto the fan inlet or built into the fan suction eye. The guide vanes must be selected for the correct handing of the fan so that the vanes direct air onto the impeller blades. The relative power requirements are normally presented against a percentage of the design flow rate as shown in Fig. 15.14.

Speed Control

For economy of operation, speed control of a centrifugal fan is probably the most satisfactory, since fan power is proportional to the cube of fan speed.

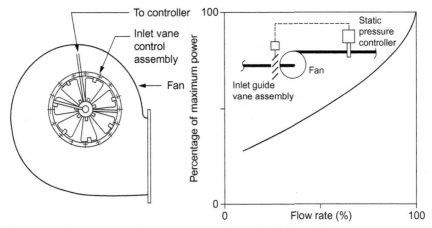

Fig. 15.14 Flow rate control of a centrifugal fan using inlet guide vanes.

This may be achieved in a number of ways. One method is by using an AC motor whose speed is regulated through a variable-frequency static inverter. The efficiency of this equipment remains high over the range of system load variations, and the equipment has good reliability. Other methods of speed control include variable pulley drive, fluid coupling, and DC motor.

SELECTING THE FAN FOR THE SYSTEM

Fan static pressure is defined by:

$$P_{sF} = P_{tF} - P_{\bar{v}F} \tag{15.8}$$

The fan velocity pressure $P_{\bar{v}F}$ is determined by using the mean velocity $\bar{v}_F$ at the fan discharge.[1] It is defined by Eq. (15.8) for two reasons:
- For fan selection when the losses at fan suction and discharge are unknown,
- To allow for the occasion when the fan velocity pressure is not available to deal with any system pressure losses.

---

[1] Fan static pressure defined by Eq. (15.8) should not be confused with the actual static pressure rise across the fan.

Initially, when calculating the total pressure loss through the system, the losses at the fan suction and discharge connections cannot be calculated, since the fan at this stage will not have been selected. The total system pressure loss may therefore exclude losses at these connections, and the fan is selected on fan static pressure. When installed, the fan will operate on the total pressure developed, and so, by definition, one fan velocity pressure will be available to deal with the losses at the fan connections. Alternatively, the losses at the fan suction and discharge connections can, in theory, now be calculated, based on the first fan selected, added to the original total and the fan *reselected* on fan total pressure (an iterative process). This is rarely done however as the pressure losses at the fan connections, especially in the discharge, are likely to be larger than those obtained from *textbook* losses, where the loss coefficients are for fully developed, turbulent pipe flow conditions.

One further point must be made. Even with one fan velocity pressure available to deal with the losses at suction and discharge connections, excessive losses can occur if a centrifugal fan is not *handed* correctly. This would also be the case if the suction and discharge connections are not designed according to good practise. It should be noted in passing that when a fan/system fails to deliver the design duty, the cause can often be traced to an unsatisfactory duct arrangement close to the fan. Some investigations have been made into what is termed the *fan/system effect* in order that any additional losses may be accounted for. These effects are similar to the interference effects between fittings described in Chapter 14.

In detailing the connections, the aims are:

- to produce a smooth, even flow into the fan suction eye;
- to ensure the fan is handed correctly in relation to the ductwork on its discharge;
- to make any expansion of the discharge duct that takes place gradually. Well-designed connections will also ensure minimum air noise generation from these fittings. Some typical poor and improved duct arrangements are illustrated in Fig. 15.15 [2].

FAN NOISE

Fans generate noise in a variety of ways, for example, broadband noise from turbulent flow over the blades and discrete tones due to the interaction between the wakes leaving the impeller blade and the vanes. This noise radiates directly into the upstream and downstream ducts. Invariably, a fan will generate least noise at a duty corresponding to maximum efficiency so that it is important to match the fan correctly to the required system design point. Where a number of fans are available for the same duty/efficiency point, the fan with the lower tip speed will generate least noise.

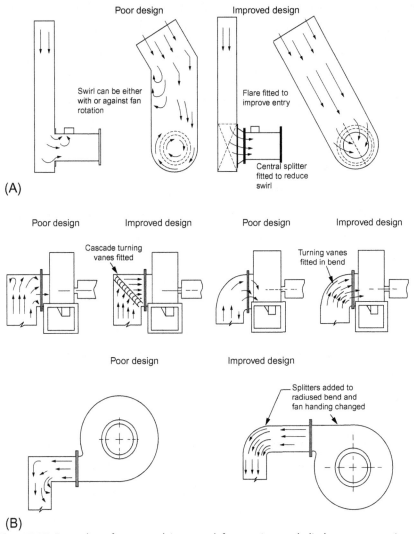

(B)

Fig. 15.15 Examples of poor and improved fan suction and discharge connections. (A) Effect of lateral bends in fan inlet ducting (axial fan shown—applies equally to centrifugal fans) and (B) bends on fan discharge. *(Figures based on Figs 9.5, 9.6, and 9.10 from Fan Application Guide, 1981, the Fan Manufacturers' Association, with permission).*

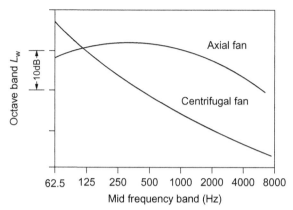

Fig. 15.16 Typical sound power level characteristics of centrifugal and axial flow fans.

The general shape of the sound power level spectrum of centrifugal and axial flow fans is given in Fig. 15.16. Subjectively, centrifugal fans appear to be less noisy to the ear than axial fans; this is because they generate more low-frequency noise. However, it is sometimes cheaper to install a higher speed, smaller diameter axial fan with an attenuator than a slower speed centrifugal.

The physical size of the fan will be related to both the duct sizing and the installation of the fan to give adequate space for maintenance, repairs, cleaning, and possible replacement, but the fan should be as large as possible to keep discharge velocities as low as possible with a recommended maximum of between 10 and 16 m/s.

With CAV systems, the flow rate may vary to some extent due to increased system resistance from such plant items as clean/dirty filters. It is therefore a good practise to select the fan with an operating point to the right of the maximum efficiency point. With a VAV fan, greater care will be needed to select a fan, together with its method of flow rate regulation, which will meet the acoustic requirements of the system.

Where the building is continuously occupied and at night lower noise levels are desirable, e.g., in hospital wards, two fans in parallel (or a two-stage axial fan) selected for the maximum duty provide the opportunity to reduce noise levels by operating only one of the fans.

Flexible Connections and Anti-vibration

Flexible connections should be used where the fan connects directly to the ductwork on the suction and discharge. Anti-vibration mounts should be used to isolate the fan from the building structure. With indirect drives, the fan and motor must be mounted on a common base with the base isolated from the structure. Care must be taken that no 'bridging' straps connect these with other anti-vibration devices.

TESTING FANS FOR GENERAL PURPOSES

BS EN ISO 5801:2008 [3] includes the important provision for site testing. Four types of fan installation are recognized:
- Type A—free inlet and free outlet,
- Type B—free inlet and ducted outlet,
- Type C—ducted inlet and free outlet,
- Type D—ducted inlet and ducted outlet.

To ensure satisfactory site measurements, the system should be designed to include the necessary recommended test lengths. Duct connections modify fan performance; consequently, a fan adaptable for more than one type of installation may have more than one standardized performance characteristic. Site installations rarely conform to standard test duct arrangements, and pressures obtained from a site test are strictly applicable to the site installation. Pressure measurements should be made as close to the fan as possible, to ensure minimum error from the pressure loss due to friction.

Site measurements of fan performance often do not correspond to the catalog data. The fan manufacturer's published characteristics are obtained under *ideal* (laboratory) conditions, and almost any *system* will modify its stated performance. In addition to any obvious duct friction losses, the effect of the duct configuration close to the fan must be considered. All site readings should be corrected to standard air conditions and fan running speed obtained at the same time as the flow rate and pressure measurements.

SYMBOLS

A', B'	constants of fan characteristic
C_{FV}	constant of proportionality for fan flow rate
C_{Fp}	constant of proportionality for fan pressure
C_{FP}	constant of proportionality for fan power
D	diameter
n	fan speed
P_a	fan airpower
P_i	fan impeller power
p_{at}	atmospheric pressure
p_v	velocity pressure

p_{sF} fan static pressure
p_{tF} fan total pressure
p_{vF} fan velocity pressure
r_t total system resistance
$\dot{V}$ volume flow rate
η fan efficiency
ρ air density

REFERENCES

[1] BS EN ISO 5801:2008, BS 848–1:2007, Industrial fans—Performance testing using standardized airways, British Standards.
[2] Fan Application Guide, HEVAC Association (UK), 1981. See also Fan Application Guide CIBSE (with HEVAC Association) TM42: 2006.
[3] BS EN ISO 5801:2008, op. cit.

CHAPTER 16

Balancing Fluid Flow Systems

As pointed out in Chapter 14, design engineers should aim to balance the fluid flow circuits at the design stage of a project so that when the systems are brought into operation, each outlet or unit will operate at the design flow rate, within specified tolerances. This is a theoretical ideal since it is almost certain that pressure losses of the installed system will differ from the calculated losses, and it will be found from initial measurements that the required balance has not been achieved. It is therefore necessary to incorporate devices for measuring and regulating the flow in the duct and pipe systems, to allow on-site balancing. Even with systems that are considered to be self-balancing, it may still be necessary to include flow measurement stations to verify system performance and to allow investigations into the reasons for any apparent failure of the system to deliver its rated output (see Chapter 19 for system testing). An understanding of on-site balancing procedures is essential so that the design engineer can plan the distribution runs of pipes and ducts and locate the regulating devices in the correct positions to aid those procedures.

On-site balancing of a fluid flow system can be defined as the adjustment of the flow rates to correspond to those specified in the design. The reasons for measuring and balancing are:

- for air systems to achieve:
  - o the design ventilation rate;
  - o the required air movement within the space or building.
- for water systems, to ensure that heat transfer equipment is supplied with design flow rates within specified tolerances;
- to provide a basis for assessing the ability of the system to maintain design conditions of temperature and humidity;
- to allow accurate testing of plant components;
- to ensure efficiency of operation: an unbalanced system is likely to work at a lower overall efficiency than a balanced one.

THEORY OF PROPORTIONAL BALANCING

Two methods of balancing have been used with success. Both methods are known as proportional methods, the absolute (design) flow rate, within the

Air Conditioning System Design
http://dx.doi.org/10.1016/B978-0-08-101123-2.00016-9

required tolerances, being obtained at the end of the balancing procedure. The methods are based on the following theory.

Consider a duct (or pipe) **WX** supplying two branches, 1 and 2, with volume flow rates $\dot{V}_1$ and $\dot{V}_2$ as shown in Fig. 16.1. For a pipe system, there will be a return **YZ**, whilst for an air system it will be usual for both flow rates to discharge to a common datum, close to atmospheric pressure.

Using the concept of system resistance for fully developed turbulent pipe flow (Eq. 13.16):

$$\Delta p = r\,\dot{V}^2$$

Branches 1 and 2 have resistances r_1 and r_2. The pressure drop from **X** to **Y** is the same for both branches; therefore:

$$\Delta p_1 = \Delta p_2$$

$$\therefore r_1\,\dot{V}_1{}^2 = r_2\,\dot{V}_2{}^2$$

$$\therefore \frac{\dot{V}_1}{\dot{V}_2} = \sqrt{\frac{r_2}{r_1}} = \text{constant}$$

Assuming the friction factor and velocity pressure coefficients remain constant and with no further damper or valve adjustment, the ratio of the flow rates $\dot{V}_1/\dot{V}_2$ will therefore remain constant irrespective of any change in the total flow rate $\dot{V}_t$ in **WX**. In terms of a balancing procedure, this means that the outlets on any branch that are furthest away from the fan/pump must first be adjusted so that the ratio of the *measured volume flow rate* $\dot{V}_M$ to the *design volume flow rate* $\dot{V}_D$ is equal. This ratio, in balancing parlance, is known as the R ratio, i.e.,

$$R = \frac{\dot{V}_m}{\dot{V}_d} \tag{16.1}$$

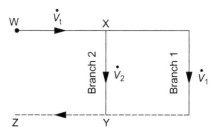

Fig. 16.1 Flow network, proportional balancing theory.

When R ratios are equal, the individual flow rates from the outlets being compared are in the same proportion to one another. Then, when the total flow rate is adjusted to the design flow rate, the balanced flows will be maintained, with all outlets delivering their design flow rates.

To illustrate these principles, consider the three outlets on one branch in the simple ducted air system in Fig. 16.2.

Each outlet has the same design flow rate, but when the fan is operating with all the dampers fully open, the actual flow rates do not correspond to the design flow rates, so the system requires balancing. Outlet 1 is obviously the *index* outlet since it has the lowest measured flow rate compared with the design, and its damper will remain fully open. It should also be noted that the R ratio of 0.6 at outlet 1 is the lowest of the three outlets. This R ratio indicates that the outlet is delivering 40% less air than design. When the damper in outlet 2 is adjusted to give the same measured flows in both outlets 1 and 2, these two will be in proportional balance, though still not delivering their design flow rates. Note that the flow rate in outlet 1 will change as the damper for outlet 2 is adjusted since this will increase the resistance of branch 2. Therefore, each time damper 2 is altered, a comparison of both flow rates must be made. Moving toward the fan, the damper for outlet 3 should now be adjusted to bring its flow rate into proportional balance with the other two, but in doing this, the flow rate in **XY** will increase, thus changing the flow rates from outlets 1 and 2, though not upsetting their *balance*. Again, it is important to note that the reduced flow in outlet 3 must be compared with the increased flow in outlet 1 until their R ratios correspond. When this proportional balancing is complete, the total flow delivered by the fan is adjusted to give the design flows from the three outlets. Once more, the

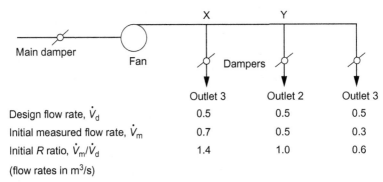

	Outlet 3	Outlet 2	Outlet 3
Design flow rate, $\dot{V}_d$	0.5	0.5	0.5
Initial measured flow rate, $\dot{V}_m$	0.7	0.5	0.3
Initial R ratio, $\dot{V}_m/\dot{V}_d$	1.4	1.0	0.6
(flow rates in m³/s)			

Fig. 16.2 Simple system to illustrate proportional balancing procedures.

change in total flow will not upset the balance of the three outlets, and each will deliver the design flow rates within the tolerance of the balance and within the accuracy of the measurement technique and instrumentation employed.

Two methods of proportional balancing, which are described in greater detail below, refer to the *index* outlet (in some literature this is termed *the least favoured* outlet). This is the outlet having the smallest R value in any group of outlets served by a branch duct before the balancing starts and when all dampers/valves are fully open. Ideally, the damper/valve on the index branch will be fully opened after the balancing procedure is complete. With the CIBSE Commissioning Code A method [1], this is not always easy to achieve, e.g., when the end outlet is not the *index*, whereas the Ma method [2] easily overcomes this difficulty.

Preliminary Checks and Adjustments

- The system must be capable of being balanced, e.g., the system should include dampers/valves in suitable positions, and it must be possible to measure the flow rates with an accuracy consistent with the required tolerance of balance.
- The information required includes drawings of plant and ductwork/pipework layouts. A simple line diagram with all outlets labelled will be of considerable help; such a diagram will include the design flow rates.
- Check that, for air systems, the plant sections and ductwork are clean and airtight, that the fan or pump is running correctly, and that there are no blockages. Wet systems should be flushed through, refilled, and vented.
- All outlet devices and fittings should be in place, and all dampers/valves must be fully opened on the branch being balanced. It is also necessary to have any associated systems, such as air extract and running.

Having taken these precautions and adjusted the fan/pump to pass approximately the total design volume flow, everything is ready for the balancing procedure to commence.

The detailed procedures and examples given below are for ducted air supply systems, but the same principles also apply to extract air systems and water systems. The on-site balancing process is performed within a stated *tolerance of balance* (see Table 16.3), though for illustration purposes in the given examples approximately equal values of R are quoted at the end of each stage in the procedure.

CIBSE Commissioning Code

The method of proportional balancing in the CIBSE Code A for air distribution systems is described below. In many ways, this method is the most attractive procedure for air systems because it requires no special equipment or fittings. Calibrated dampers are not required (unlike the Ma method described later), and the method is generally appropriate to systems that are already designed and installed. A suitable anemometer is required for measurements at the outlets, with in-duct measurement facilities being preferred in the main branches.

The procedure listed below is illustrated in Table 16.1 where a main duct supplies five outlets:

Table 16.1 Example of CIBSE method of proportional balancing

Damper					
Outlet no.	5	4	3	2	1
Design flow rate, $\dot{V}_d$	0.11	0.14	0.17	0.15	0.10

Outlet 1 is found to be the index from initial measurements

Measured flow rates at outlets 1 and 2 $\dot{v}_m\left(m^3/s\right)$				0.18	0.09
$R = \dot{v}_m/\dot{v}_d$				1.20	0.90
Adjust damper 2 to give, after a few trials, equal R ratios				0.98	0.98

Measure flow rate at outlet 3 $\dot{v}_m\left(m^3/s\right)$			0.23	–	–
$R = \dot{v}_m/\dot{v}_d$			1.35	–	0.98
Adjust damper 3 to give, after a few trials, equal R ratios			1.05	(1.05)	1.05

Measure flow rate at outlet 4 $\dot{v}_m\left(m^3/s\right)$		0.21	–	–	–
$R = \dot{v}_m/\dot{v}_d$		1.50	–	–	1.05
Adjust damper 4 to give, after a few trials, equal R ratios		1.10	(1.10)	(1.10)	1.10

Measure flow rate at outlet 5 $\dot{v}_m\left(m^3/s\right)$	0.19	–	–	–	–
$R = \dot{v}_m/\dot{v}_d$	1.73	–	–	–	1.10
Adjust damper 5 to give, after a few trials, equal R ratios	1.13	(1.13)	(1.13)	(1.13)	1.13

Notes: (1) After adjusting damper 5, all outlets are delivering 13% more than their respective design flow rate. If this was the only branch of the ducted air system, the total flow rate would be reduced to deliver the design flow rate of $0.67 \text{ m}^3/\text{s}$. (2) The bracketed figures indicate flow rates that are not measured but remain in proportional balance. It is unlikely that equality of R ratios will be achieved with the first adjustment of a damper and a number of trials may have to be made.

(a) Find the *index* outlet by inspection or measurement.
(b) If the *index is* not the end outlet (1 or 2), reduce the flow in the end outlet to make it so by reducing its flow rate.
(c) Measure the flow rates in outlets 1 and 2.
(d) Calculate the R ratios for outlets 1 and 2, i.e.,

$$R = \frac{\text{measured volume flow rate}}{\text{design volume flow rate}} = \frac{\dot{V}_m}{\dot{V}_d}$$

(e) Adjust damper 2 to give the same value of R for both outlets 1 and 2.
(f) Following each adjustment, measure the flow rate of the index outlet and recalculate values of R. It is unlikely that equality will be achieved with the first adjustment and a number of trials may have to be made.
(g) Measure the flow rate at outlet 3 and calculate values of R.
(h) Adjust damper 3 until the value of R is the same for both outlets 1 and 3.
(i) Repeat steps (f) and (g) for as many outlets as are supplied by the branch duct.

Multi-branch Systems

A more complex system is now considered with a number of branches each supplying a group of outlets, as shown in Fig. 16.3.

In this case, each group of outlets on individual branches is brought into proportional balance independently of the other branches. Each branch will then have a different R ratio, e.g., the R values could be:

$$R_a = 0.7, \quad R_b = 0.9, \quad R_c = 1.2, \quad \text{and} \quad R_d = 1.4$$

The requirement is now to balance these four branches using the procedure already described. In this example, branch **B** is balanced to branch **A** (the index), then branch **C** to **A**, and finally branch **D** to **A**. When this has been completed, the total flow in the system will be adjusted to the design flow rate. This example illustrates the necessity for the designer to include dampers in the branches before the first outlets, i.e., in branches **B, C,** and **D**.

These procedures can be followed through with any complexity of circuit, provided that the principles are understood both by the system designer and by the commissioning engineer.

Effect of Branch Balancing Dampers

Balancing procedures rely on the static pressure gradients remaining uniform along the duct when flow rates are either reduced or increased. When a damper is in a partially closed position, the pressure distribution along the

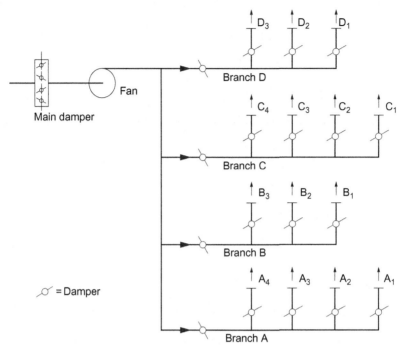

Fig. 16.3 Balancing a multibranch network.

duct, downstream of the damper, has the same general characteristics as that of an orifice plate—that is, there will be static depression followed by pressure recovery. If the first offtake on a supply branch duct is too close to the balancing damper, then the proportional balance of the downstream outlets will be affected when that damper is closed. To avoid this problem, the first offtake on a branch duct should be a minimum of two to three duct diameters downstream of the branch damper.

Last Outlet Not the Index

Consider the branch of a system with the prebalanced flow rates shown in Fig. 16.4. Here, the index outlet is identified as no. 4, which is *closest to the* fan in the sequence in which the ducts are taken off the main duct. The reasons for this outlet being the index are either that it has the longest duct run (the figure is diagrammatic) or that fittings have higher pressure drops than the other branch ducts. If the balancing proceeds from outlets 1 and 2, without first identifying the index as outlet 4, then, when the commissioning engineer reaches that outlet, balance could not be achieved without back

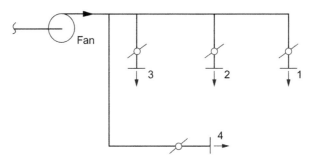

Fig. 16.4 Balancing a branch when the end outlet is not the index outlet.

tracking, something that good procedures seek to avoid. Hence, step (b) of the procedure is to identify the index outlet by measurement. Then, if the end branch (either 1 or 2) is not the index, its damper is adjusted to make it an *artificial* index. Experience is required in making a suitable assessment on the amount of reduction—the CIBSE Code suggests making it equal to the actual index. The expectation is that, when the operator reaches outlet 4, the damper should be fully opened. If the original judgment of closing damper was incorrect, too much closure should mean that either damper 4 should have to be closed or the procedure would have to be repeated. At the end of balancing, the actual index outlet should be fully opened. If it is not, there may be penalty on fan energy consumption. There also may be unwanted noise generated by the other dampers since they would be closed more than they need be.

Two design points are illustrated by this. First, the end branch should have a damper (even if the designer has taken it to be the index outlet). Second, systems should be designed to assist the balancing procedures. In this particular example, the design engineer should try to design the system in such a way as to make the last outlet the index so that the problem does not arise. There are, of course, other practical and theoretical considerations to be taken into account when designing the circuits. The implications of commissioning and balancing are only part of the overall design considerations and must be reconciled with other requirements to obtain an optimum design.

Ma's METHOD

The method of balancing developed by Ma makes use of dampers that have a calibration relating damper position to reduction in flow rate. Since the reduction in flow rate in any outlet has been related to the index outlet, this calibration will be in terms of a proportional flow rate ratio, i.e., a ratio F is calculated as:

$$F = \frac{R \text{ ratio obtained in the index outlet}}{R \text{ ratio obtained in outlet to be regulated}} = \frac{R_i}{R_o}$$

For circular ducts, a suitable, commercially available, damper is the iris shutter type for which the calibration curve in Fig. 16.5 was obtained over a wide range of damper sizes (100–350 mm diameter) and air velocities (3–7 m/s). For rectangular ducts, an opposed blade damper would be suitable, provided that an appropriate calibration was available.

Taking any one branch with a group of outlets, the procedure is to set all dampers, in one operation, relative to the index. Because of the complexities of the pressure changes within the system, the initial settings are unlikely to give a balance within required tolerances. The measurements have therefore usually to be repeated (and finer adjustments made) to give a more precise balance. In other words, it is an iterative procedure that continues until balance has been achieved to the specified tolerances; in practise, only one or two repetitions will be required.

After the initial preliminary checks, the balancing procedure listed below is illustrated in Table 16.2:

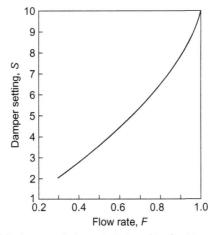

Fig. 16.5 Calibrated iris damper characteristic (used in the Ma method of proportional balancing).

Table 16.2 Example of the MA method of balancing using calibrated dampers

	Damper			
Outlet no.	4	3	2	1
Design flow rate, $\dot{V}_d$	0.35	0.25	0.20	0.30
Stage 1: initial measurements				
Measured flow rates, $\dot{v}_m \, (m^3/s)$	0.40	0.27	0.16	0.20
$R = \dot{v}_m / \dot{v}_d$	1.14	1.08	0.80	0.67
$F = R_i / R_o$	0.59	0.62	0.83	1.00
Damper setting, S (Fig. 15.5)	4.4	4.7	6.8	10.0
Adjust dampers to setting S				
Stage 2: first iteration				
Measure flow rates, $\dot{v}_m \, (m^3/s)$	0.31	0.20	0.18	0.25
$R = \dot{v}_m / \dot{v}_d$	0.89	0.80	0.90	0.83
$F' = R_i / R_o$	0.93	1.04	0.92	1.00
Damper setting, $S' = S \times F'$	4.1	4.9	6.3	10.0
Adjust dampers to setting S'				
Stage 3: final measurements				
Measure flow rates, $\dot{v}_m \, (m^3/s)$	0.30	0.21	0.17	0.26
$R = \dot{v}_m / \dot{v}_d$	0.86	0.84	0.85	0.87

Notes. (a) Outlet 1 is the index, and its damper remains open throughout; (b) after stage 3, each outlet is delivering approximately 85% of the design flow rate; if this was a single branch of the system, the flow rate would be increased to deliver the design flow rate.

Stage 1

(a) Measure air flow rate to each outlet.

(b) Calculate the R ratio for all outlets:

$$R = \frac{\dot{V}_M}{\dot{V}_D}$$

(c) Divide lowest R (R_o) value (the *index* outlet R_i ratio) by the other ratios. This gives the proportion by which rate of airflow must be reduced, i.e.,

$$F = \frac{R_i}{R_o}$$

(d) Enter the F ratios calculated in step (c), individually into the calibration curve to find damper setting S (Fig. 16.5).

(e) Adjust damper to setting S.

Stage 2 (first iteration)

    (f) Measure air flow rates to each outlet and calculate new R values.

    (g) Obtain new values for R_i divided by values of R_o $(=F')$. Note that the R_i value is for the *original* index outlet.

    (h) Multiply ratios F' by damper settings obtained in step (d):

$$S' = F' \times S$$

    (i) Adjust damper to new setting S'.

Stage 3 (second iteration, if required)

    (j) Measure air flow rates and, if necessary, proceed to a third adjustment following steps (f) to (j).

Stage 4

    (k) Adjust the main balancing damper or the fan speed to obtain the design air flow rate.

Where systems have more than one branch, each branch is treated independently and afterward proportionally balanced to each other. Finally, the fan total volume flow is adjusted to give design flow rates from each outlet in a similar manner to that described above for the system shown in Fig. 16.3.

One of the advantages of the Ma method of balance is that it is not dependent on the end branch being the index outlet. If the procedure is applied correctly, the damper on the index will remain fully open.

TOLERANCE OF AIR FLOW RATE BALANCE

Neither it is unnecessary and impractical to obtain identical R values for all outlets or branches within a group, nor is it necessary to regulate the total flow to an accurate design total air flow rate. Realistic tolerances should be established so that balance can be achieved with economy of effort. The tolerances suggested in the CIBSE Code A are given in Table 16.3.

Both the methods described above are capable of achieving balanced systems within the recommended tolerances.

MEASUREMENT OF AIR FLOW RATES AND BALANCING

Measurement at the Outlets

In many ducted air systems, a vane anemometer is the most convenient instrument with which to make measurements at the outlets. In the CIBSE Code, the recommended method for measuring the mean velocity $\bar{v}_a$ at the face of a grille is given below. The air volume flow rate is then calculated as:

Table 16.3 Tolerances on air system balance

Type of system	Terminals	Branches	Total air flow
Mechanical ventilation and comfort cooling	+20% of lowest terminal	+10% of lowest terminal	From +10% to +5%
Air conditioning and pressurization escape routes	+15% of lowest terminal	+8% of lowest terminal	From +10% to 0%
Close control air conditioning	+10% of lowest terminal	+5% of lowest terminal	From +5% to +0%

$$\dot{V} = \bar{v}_a A_g \tag{16.2}$$

where A_g is the gross area of the grille.

Provided that the outlets are similar, it is usually unnecessary to consider the effect that the *free area* has on this calculation, when the outlets are being proportionally balanced. This is because the R ratios only are being compared and any *grille effect factor* C_g is considered to be constant. When an accurate flow rate is required from the measurement, then the measured flow rate must be multiplied by the grille effect factor, which is either supplied by the manufacturer or obtained from a laboratory calibration. The flow rate is then obtained from:

$$\dot{V} = C_g \bar{v}_a A_g \tag{16.3}$$

From laboratory tests, C_g is of the order of 0.85 for supply grilles and 0.90 for extract grilles, where the grille sizes are above approximately 0.1 m^2 [3]. Grille free areas should *not* be used as grille effect factors.

Method for Measuring the Mean Velocity

The grille face is divided into, say 6, equal areas, as in Fig. 16.6.
 · The mean face velocity is obtained as follows:
(1) Check direction of flow through instrument.
(2) Position the anemometer in the centre of each rectangle in turn, with the back of the instrument touching the front of grille; for each rectangle record, the indicated velocity, v_i. If there is needle fluctuation, try to judge the mean position.
(3) Calculate the arithmetic mean of all 6 values of v_i, to obtain $\bar{v}_i$.
(4) Correct $\bar{v}_i$ according to instrument calibration to give $\bar{v}_a$.
(5) Note that an average correction factor is sufficiently accurate for site measurements.

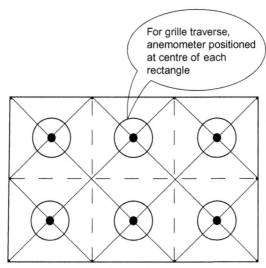

Fig. 16.6 Anemometer traverse positions at grille face.

Effect of Balancing Damper on Grille Face Measurements

With many grilles, the balancing damper is often incorporated as a composite unit, immediately behind the blades of the grille. To balance the air quantities, it will be necessary partially to close the damper to provide the *out-of-balance* resistance in relation to the index outlet. When a damper is partially closed, the higher velocity jets strike the anemometer vanes causing the instrument to indicate an average velocity higher than with the damper fully open, at the same flow rate. The grille effect factor then varies considerably, and if not taken into account, accuracy of flow rate measurement becomes much greater than the required tolerance against which the system is being balanced. The typical variation of the grille effect factor C_g with the damper position is given in Fig. 16.7.

Alternatives should therefore be considered to overcome this problem of comparative accuracy between flow measurements at supply grilles; these include the following:

- *A hood-mounted anemometer.*

    The advantage of this method is that the velocity only needs to be recorded and the R ratio calculated as $(v_a / \dot{V}_d)$. A disadvantage is that the additional resistance imposed by the hood will introduce an

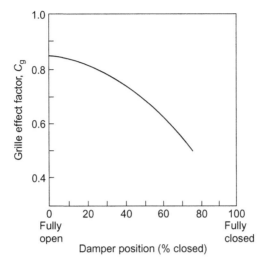

Fig. 16.7 Grille effect factor on anemometer reading.

additional uncertainty into the accuracy of balance. Also, a number of different sizes of hood might be required for different sizes of outlet.

- *Measure the flow rate upstream of the damper.*

    For this, access to the ductwork must be provided at the design stage of the project.

- *Separate the damper from the grille with a length of duct.*

    Again, this is a solution that must be incorporated into the original design; it has the additional advantage of allowing acoustic treatment of the duct where air noise generated by the damper is likely to cause annoyance.

- *Use an extension duct.*

    Here, an instrument traverse is made away from the grille face. This is not often practical since several extension pieces may be required.

Measurements in extract systems are unaffected by a damper immediately behind the grille.

Anemometers mounted in a hood are recommended when measurements are made at supply diffusers (Fig. 16.8).

In-Duct Measurements

Balancing in a multibranch system (as illustrated in Fig. 16.3) can be achieved by using the index outlets on each branch, as representative

Fig. 16.8 Hooded vane anemometer. *(Courtesy of Inlec UK Ltd.)*

individual flow rates. However, it is more accurate and often more convenient to make in-duct measurements close and upstream of the dampers to be adjusted. The traverse methods of obtaining the flow rate in a duct have been described in Chapter 13. For air velocities above 3.5 m/s, it will be usual to use Pitot-static tubes, and for velocities below 3.5 m/s, either a small diameter vane anemometer or a hot-wire instrument should be used.

Flow Rate Using a Centre Constant

If in-duct measurements are made to bring a group of branches into proportional balance, repeated traverses become time-consuming. Single-point measurements to obtain a flow rate are particularly useful when repeated measurements have to be made in the same duct, as in proportional balancing procedures. Unless the duct flow is fully developed with a symmetrical velocity profile (and outside the laboratory it rarely is), it is not possible to rely on published centre constants since the installed duct lengths are relatively short between fittings and the velocity profiles will be asymmetrical.

To obtain a centre constant C_c requires making a preliminary traverse of the duct to determine the ratio of the mean to axial velocities, i.e.,

$$C_c = \frac{\bar{v}}{v_c} \tag{16.4}$$

The centre constant is then multiplied by the centre point measurement of velocity to determine the flow rate, i.e.,

$$\dot{V} = C_c\, v_c A_d \qquad (16.5)$$

This method of obtaining the flow rate would not be suitable if the velocity distribution in the duct changed due to varying inlet conditions, e.g., from a balancing damper upstream of the measuring station. (For this reason it is important to have the measuring position upstream of the damper.) It is also limited by the accuracy of the centre constant and the accuracy of the subsequent centre point velocity measurement.

BALANCING WATER SYSTEMS

As with ducted air systems, it should be the aim of the design engineer to balance water circuits at the design stage of a project. Where this is not possible, it will be necessary to include devices for regulating and measuring the flow rates in the piping circuits to allow on-site balancing.

Most plant items require isolating valves for maintenance. Traditionally, the valve in the return pipe has also had a regulating function. Current balancing techniques use both valves for flow measurement.

Measurement of Water Flow Rate and Balancing

The flow rate through an orifice plate is given by Eq. (13.20). Although the flow coefficient α in this equation depends to some extent on Reynolds number, these variations may be discounted for site balancing procedures. The expansibility factor may be taken as constant, at $\in\, = 1$. If the density of water is also assumed to be constant, then, for a fixed-orifice-to-pipe-diameter ratio, Eq. (12.20) can be written as:

$$\dot{V} = K_{vs}\sqrt{\Delta p'} \qquad (16.6)$$

Thus, the characteristic of a fixed-orifice flow-measuring device may be expressed by a single constant K_{vs}.

The proportional balancing procedures compare the ratios of *measured* to *design* flow rate (the R ratio) and from Eq. 16.1, it then follows that:

$$R = \frac{\dot{V}_m}{\dot{V}_d} = \sqrt{\frac{\Delta p'_m}{\Delta p'_d}} \qquad (16.7)$$

Therefore, when fixed-orifice measuring devices are used for measuring the flow rates, the pressure difference ratio only is required for balancing purposes, the valve flow rate constant K_{vs} being used to determine the design pressure difference, $\Delta p_d{}'$.

The CIBSE Code W [4] recommends that hot water systems are balanced cold, which for high-temperature water systems is important for safety reasons. If measurements of hot water flow rates are made, the correction to Eq. (16.6) due to water density variations will be $\pm\sqrt{1000/\rho}$. For water at 100°C, this correction amounts to approximately $\pm2\%$. Proportional balancing assumes that the circuit resistances obey the *square law*, i.e., both the friction factors and the fitting pressure loss coefficients remain constant. In fact, both depend, to some extent, on Reynolds number, and this means that the system balance will alter when the total flow rate and the water temperature change. Consequently, the tolerance against which the system is balanced should take account of the possible variations in Reynolds number.

The engineer will be aware that the pressure difference for measuring the flow rate is not the same as the net pressure drop. This point is illustrated in Figs 13.20 and 13.21 in Chapter 13. The downstream pressure tapping sits in the low-pressure region, and there is a pressure recovery after the device, before the net pressure loss is obtained. It is important to remember this when referring to tables and charts published by manufacturers who sometimes use terms incorrectly. At least one manufacturer uses the term *signal* to describe the measured pressure difference, and the adoption of this term may lead to less confusion in the presentation of information.

Flow Measurement Valve

Flow measurement valves may be an orifice valve or a flow-measuring device (FMD) such as an orifice plate that is close coupled to a gate valve shown in Fig. 16.9. Manufacturers provide the flow rate constants, K_{vs}, manufacturers also present the pressure difference/flow rate characteristics as log-log graphs as shown in Fig. 16.10.

Double Regulating Valve

A double regulating valve (DRV), shown in Fig. 16.10, is used for regulation and isolation. Once a branch has been regulated (brought into proportional balance), the valve can be locked off to prevent further opening. Then, if the

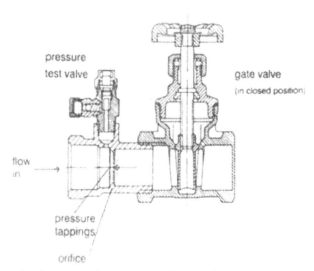

Fig. 16.9 Fixed-orifice gate valve. *(Courtesy of Crane Fluid Systems)*

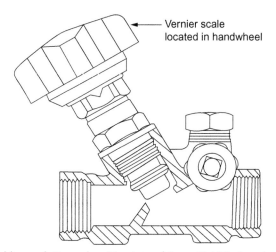

Fig. 16.10 Double regulating valve. *(Courtesy of Crane Fluid Systems.)*

valve has to be closed to isolate the circuit, when reopened, it will return to its original setting. Most of these valves include a vernier scale on the stem to allow accurate setting. Hand wheels can be removed to reduce the risk of unauthorized tampering.

A DRV will usually be placed in the return pipe and used in conjunction with a FMD installed in the flow pipe. When used in this way, there is no need for a DRV to include pressure tappings as in the variable orifice double regulating valve described below. A DRV is sometimes close coupled with an orifice plate to form a fixed orifice double regulating valve (FODRV). This arrangement is sometimes referred to as a *single-valve commissioning set*, its rationale being that the engineer can more readily observe the change in pressure difference as the valve is adjusted.

Variable Orifice Double Regulating Valve

A variable orifice double regulating (VODRV) is a DRV with pressure tappings. The valve has the functions of the DRV, as well as the ability to measure the flow rate. The flow rate/pressure difference characteristics are usually presented graphically on a log-log graph (Fig. 16.11). For reasonable accuracy of measurement, a vernier scale for setting the valve is essential.

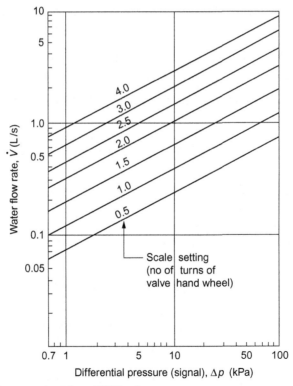

Fig. 16.11 Flow rate chart for a FODR valve.

Where VODRVs are used for regulating and measuring the flow rates, it will be necessary to determine the flow rate for the R ratio, rather than using the differential pressures, since K_{vs} varies with valve position.

Nozzles, Orifice Plates, and Venturi Metres

Orifice plates, nozzles, and venture metres are marketed as either flow-*measuring devices* or *metering stations*. Orifice plates are often close-coupled to a gate valve or to a DRV, thus forming a composite unit as previously described. When close-coupled, the static pressure distribution through the unit may be affected by the valve, and the flow coefficient α will then differ from those published in the ISO Standard ISO 5167-1: 2003 [5]. In these cases, the pressure difference/flow rate characteristics given by the manufacturer should be used, not ISO the published values.

Pressure Tappings and Pressure Test Valves

Pressure tappings allow a manometer to be connected to the flow-measuring device. These devices are self-sealing and usually suitable for cold and low temperature hot water (LTHW) systems. The manometer tubes incorporate matching self-sealing connectors; these may be push-on units for quick connect/disconnect to the system, but more usually, a threaded fitting is used with a cap to protect the pressure tappings from dust.

A pressure test valve (PTV) is a self-sealing ball unit that also incorporates a needle valve operated by a standard aircock key. This device is therefore safer to use where the pipe fluid is at high pressure or temperature and allows the pressure tapping to be isolated for cleaning the ball seat. Where measurements are to be made on a live high-temperature hot water system, copper bleed tubes should be taken from the PTV, and they should terminate at the pipe with a needle valve. Both items (as a composite unit) are illustrated in Fig. 16.12.

Selection of Valves and Flow Measuring Devices

Balancing valves and flow-measuring devices will impose additional system resistances on the circuit. Therefore, selection should be made to ensure that these are kept to a minimum whilst ensuring sufficient pressure difference to allow accurate measurement of the flow rate. The selection should be considered at the same time as sizing the pipe since it is desirable that both should be the same size to avoid using reducers and/or expanders.

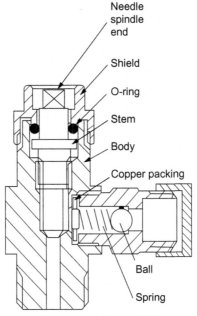

Fig. 16.12 Pressure test valve. *(Courtesy of Crane Fluid Systems.)*

The net pressure loss across a valve or an FMD is obtained from the coefficient supplied by the manufacturer.

A representative piping circuit is shown in Fig. 16.13 with an associated table giving alternative valve combinations, selected from the different devices described. The procedure for balancing the circuit will be similar to that described for the air systems. To achieve this, the control valves are initially fully open to load. When design flow rates have been obtained within the required tolerances, each of the bypass valves will then be regulated in turn, with the control valve closed to load, using the same flow-measuring device used for the circuit balancing.

Installation

The flow-measuring device, including flow measurement valves, should be positioned upstream of any associated valve and installed with a reasonable length of straight pipe upstream and downstream of the device. Where a measuring accuracy of 5% is sufficient, these lengths should be a minimum of ten and three equivalent pipe diameters, respectively. Standard lengths recommended in various standards should be adhered to if a more stringent measuring accuracy is required.

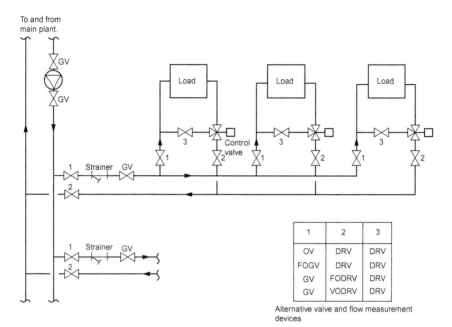

Fig. 16.13 Representative pipework circuit.

When coupling FMDs to the pipe, burrs should be removed from pipe fittings in the vicinity of the device. This will reduce the risk of creating a flow disturbance in the fluid entering the device and ensure that accuracy is maintained. Tappings should be placed either on the top or on the side of the device (or pipe) to reduce the risk of dirt accumulating in the pressure leads.

Manometers

The three types of manometer used for pressure measurements are:
*   liquid;
*   electronic digital;
*   diaphragm.

All manometers will require an isolating and equalizing (bypass) valve manifold. This is used to allow the instrument to be zeroed and purged of air.

An electronic digital manometer may have interchangeable transducers of different ranges, each of which connects to the same display unit. It will be necessary to zero the instrument for each range and each time it is used. It is desirable that the instrument is recalibrated on a regular basis. Instruments are available with direct readout in flow rate, programmable for different flow measurement devices, that is, for the instrument to accept the appropriate value of the flow coefficient C_v.

Liquid manometers use either water over mercury or water over fluorocarbon, with ranges typically 0–60 and 0–4.6 kPa, respectively. These instruments do not have to be recalibrated. For proportional balancing, it is a good practise to have a pair of compatible manometers. One of these will be positioned at the index (least favoured or reference branch), whilst the other is used as a *wander* for the units that are being proportionally balanced to the index. Tubes from the manometer must terminate in self-sealing connectors, which match the pressure tappings of the flow-measuring device. Tubing should be suitable for the temperature and pressure of the pipe fluid.

Tolerance of Water Flow Rate Balance

It is not necessary to balance flow rates within a group of loads precisely. The allowable tolerances suggested in the CIBSE Code W are from 0% to +10% for flow rates above 0.015 l/s [6]. It should be noted that, to guarantee a tolerance of balance, the accuracy of measuring flow rates should be considerably less than the tolerance of balance required by the specification.

SYMBOLS

A cross-sectional area of duct
C_c centre constant of velocity traverse
C_g grille effect factor, applied to grille face velocity
F proportional balancing ratio
K_{vs} flow-measuring device constant
R proportional balancing ratio
r resistance
S setting of a calibrated damper
$\dot{V}$ volume flow rate
v point velocity
$\bar{v}$ mean velocity
v_c centre point velocity
$\Delta p'$ pressure difference
ρ density

SUBSCRIPTS

a   vane anemometer

m   measured

d   design

i   index outlet

o   outlet other than index

ABBREVIATIONS

DRV   double regulating valve
FMD   flow measuring device
FOGV   fixed orifice gate valve
PTV   pressure test valve
VODRV   variable orifice double regulating valve

REFERENCES

[1] CIBSE Commissioning Code: Series A, Air Distribution Systems, Chartered Institution of Building Services Engineers, 2006.

[2] W.Y.L. Ma, The averaging pressure tubes flow-meter for the measurement of air flow in ventilating ducts and the balancing of air flow circuits in ventilating systems, J. Inst. Heat. Ventil. Eng. 34 (1967).

[3] R.C. Legg, The measurement of air flow at the face of a grille, vol. 47, Building Services Engineer, 1976.

[4] CIBSE Commissioning Code W, Water Distribution Systems, Chartered Institution of Building Services Engineers, 2010.

[5] ISO 5167-1: 2003 Measurement of fluid flow by means of pressure differential devices in circular cross-section conduits running full—Part 1: General principles and requirements.

[6] CIBSE Commissioning Code W, op, cit.

CHAPTER 17

Controls, Dampers, and Valves

Air conditioning systems are installed with a maximum capacity determined from indoor and outdoor design conditions and design loads, but it is only on comparatively rare occasions that these conditions require full plant capacity. Also due to the variations in external climate, building heating and cooling loads, and building use, it is very unlikely that a building air conditioning system would operate in the same mode at all times. Therefore, controls are applied to maintain appropriate indoor conditions with efficient plant capacity reduction whilst also ensuring safe and consistent operation.

OVERVIEW OF CONTROL SYSTEMS

Tim Dwyer
UCL Institute for Environmental Design and Engineering (IEDE), London, United Kingdom

There are many published works describing control systems design, both on a theoretical and practical basis. However, dampers and valves form only one part of the complete control arrangement, and before going on to discuss them in detail, a brief description of the overall area of controls is necessary.

A control system is designed to provide a variety of functions that include:
- sustaining appropriate indoor conditions,
- providing optimum running conditions for both efficient system performance and minimum maintenance,
- monitoring of space and system status,
- simplifying end-user control,
- ensuring safe plant operation.

Whatever its function, the control system comprises three basic components: *sensor*, *controller*, and *correcting unit*. The correcting unit usually consists of two items, the actuator and the final control element, e.g., damper motor and damper blades. The controller and the actuator are sometimes combined in a single unit, as in the case of a self-acting thermostatic radiator valve.

Air Conditioning System Design
http://dx.doi.org/10.1016/B978-0-08-101123-2.00017-0

Sensing Elements

The sensor is the component that feeds information to the controller relating to a variation in a property of the observed medium, which for air conditioning systems is usually water or air. Whatever medium that is being measured, the sensors need to have good sensitivity over the required range, short response time, small hysteresis (the swing in output as the condition varies up and down), resistance to contaminants (such as dust), and linear response (so that the output can be readily interpreted by the control system) and should have a maintainable long life with good repeatability of measurement and preferably be low cost.

The scope of sensors applied in air conditioning is continually broadening, but those that are more commonly used are to monitor temperature, humidity, carbon dioxide, pressure, and flow rate.

Temperature sensors are used for monitoring dry- and wet-bulb temperatures. Dry-bulb sensors include the thermocouple, thermistor, thin metal film, bimetallic elements, and fluid-filled bellows. The 'thermostat' is a particular type of temperature measuring device that, as well as sensing the temperature (using a bimetallic strip or bellows), applied the physical movement of the sensing device as the temperature changes to directly actuate a pair of contacts that may be used, for example, for on-off control.

It is rare to see a satisfactory wet-bulb sensor. However, the reliable assessment of the air dew-point temperature or moisture content is essential for air conditioning control. There is a plethora of technologies available for humidity measurement (normally providing an output related to relative humidity) including impedance (resistance or capacitance), conductivity, light, and semiconductivity. The application of the technologies in air conditioning is practically determined by the cost and availability from specific control system manufacturers.

By combing the signals from a dry-bulb sensor with that of a humidity sensor, air enthalpy may be determined to provide certain control strategies. Enthalpy control is particularly designed to reduce energy use in air conditioning such as in systems that maximize the benefits of 'free cooling' (see Chapter 5). It is particularly important that the enthalpy sensor (or combined dry-bulb/wet-bulb sensor) is set up and maintained to prevent potentially significant adverse impact of inappropriate calls for free-cooling operation.

Early CO_2 sensors were seen as unreliable and difficult to keep calibrated; however, sensors have become more reliable, less expensive, and self-calibrating. They are typically based on a nondispersive infrared (NDIR) gas sensor that uses infrared light absorption to determine the CO_2 concentration in the air.

Pressure sensors are used for both directly and indirectly measuring and controlling a variety of system conditions and are normally based on capacitive or inductive elements that use a diaphragm or bellows arrangement to pass the air pressure to the transducer

Nanotechnologies are being swiftly developed to produce sensors (and control infrastructure) across the field of building automation and air conditioning and are likely to usher in a massive shift in flexibility and usability. This heralds a change that is likely to eclipse that of the evolution of mechanical to electric to electronic sensors that has already provided immensely improved opportunities for indoor air quality, environmental, comfort, and energy sensing.

Location of the Sensing Element

The position of a sensor in the system is important in ensuring that a representative condition is observed. Temperature and humidity sensors should be placed in well-mixed airstreams, avoiding positions where there is a risk of stratification or where there is likely to be a significant temperature gradient. In cases where there is likely to be a variation of temperature across a duct, averaging sensors can usefully be employed. Temperature sensors are affected by radiant heat transfer and should therefore also be screened from high or low temperature surfaces and from the direct influence of the sun. It is particularly important that a 'dry-bulb' temperature sensor should indeed be dry and not affected by water droplets from adjacent cooling coils or humidifiers.

Pressure sensors are often applied as an indirect measure of other system parameters. This might be, for example, by providing the static and velocity pressure signals to control the minimum allowable outdoor air ventilation flow rate. They are directly employed in monitoring the static pressure in VAV systems—locating pressure sensors to reliably represent the overall pressures in a variable flow system can be challenging.

The differential pressure across an item of plant, such as a filter, provides information about its state of cleanliness for maintenance purposes. Static pressure sensors should be placed well downstream of modulating dampers (a minimum of 4–6 duct diameters) and flow-measuring sensors should be placed upstream of these devices.

The consequence of poor sensor performance may lead to excessive energy consumption, poor plant performance or, in severe cases, failure of the system itself.

The Controller

The function of the controller is to interpret the input signal from the sensor and provide appropriate output to the correcting unit. Sensors can be used to control plant items 'directly' (but still through an appropriate control mechanism) or with other elements such as resetting a thermostat that is sensing 'dew-point' air temperature leaving a cooling coil or humidifier. The action taken will depend on the design of the control action and may be undertaken in local controllers (that may be close coupled with the correcting unit), through distributed 'out stations', via centralized systems or via internet (or wide area network) connections. Although there is likely to be some processing capability locally to the sensor and actuator, it is often the building management system that will provide the overall, or overriding, control function.

The Correcting Unit

Whether considering pipework systems for liquid flow or ductwork systems for air flow, there is invariably a need for regulation of flow rates through the employment of a final control element.

The correcting unit is most likely to comprise of two parts, the actuator and the final control element.

The actuator provides the motive power to the final control element in response to a signal from the controller and may take the form of an electric or pneumatic device. The selection of the actuator type will depend on the type of final control element and on the control strategy for the overall system.

Dampers are used for regulating and controlling the flow rates in mechanical ventilation and air conditioning ductwork systems. The most common application is in the static balancing of air flow networks to the design requirements, as described in Chapter 16. Modulating dampers are often included in the central plant to allow the system to operate in an efficient manner, as in the examples of the air conditioning systems utilizing recirculated air described in Chapter 6.

For water systems, valves are employed to provide corresponding facilities to those provided by dampers in air systems, e.g., double regulating valves (DRVs) for static balancing and modulating valves for controlling the flow rates to heat exchangers such as cooler and heater batteries.

Whatever type of fluid is being regulated, there are certain parameters that apply to the control device (the damper or valve), which are common. These are used to determine the appropriate size and type of fluid flow control device (FCD).

FLOW CONTROL DEVICES

Roger Legg

AUTHORITY

The authority N indicates the ability of a flow control device (FCD) to influence the flow rate in a system. An oversized FCD would not have as much influence over the flow rate in the duct or pipe network as a smaller one. In pressure terms, the greater the pressure drop across the device when fully open, the greater the influence of the device on the flow rate through the system. The authority is defined by:

$$N = \frac{\Delta p_\varnothing}{\Delta p_s + \Delta p_\varnothing} \qquad (17.1)$$

where $\Delta p_\varnothing$ is the pressure drop in the FCD in the fully open position.

Δp_s is the pressure drop of the system excluding the FCD pressure drop.

It is important to note that the 'system pressure drop' Δp_s in Eq. (17.1) is the pressure drop in that part of the system in which the FCD controls the flow rates.

As the device closes, the pressure drop increases across the FCD, reducing the fluid flow rate. As suggested above, the authority must be significant if satisfactory control of fluid flow is required across the whole range of the device aperture. If an FCD is oversized, it will have a small authority, and control becomes effectively on/off or two-position in character.

Example 17.1

A short length of ductwork has a design pressure drop, excluding the damper, of 30 Pa. To maintain reasonable control of air flow rate, it is assumed that the damper has an authority of 0.1. Determine the pressure drop across the fully open control damper.

Solution

The system has a pressure drop of 30 Pa; rearranging Eq. (17.1), the required pressure drop across the damper in the fully open position is given by:

$$\Delta p_\varnothing = \frac{N \, \Delta p_s}{(1 - N)}$$
$$= \frac{0.1 \times 30}{(1 - 0.1)} = 3.33 \, \text{Pa}$$

INHERENT CHARACTERISTIC

The inherent characteristic γ of an FCD is defined at an authority of $N = 1$. It is the relationship between the stroke of the FCD (or angular position for damper blades) and the flow rate relative to the maximum flow in the fully open position whilst maintaining constant pressure drop across the device; i.e.,

$$\gamma = \frac{\dot{V}_\theta}{\dot{V}_\varnothing}$$

where $\dot{V}_\theta$ is the flow rate at blade angle θ,
$\dot{V}_\varnothing$ is the flow rate at blade angle $\varnothing$.

It is then shown that:

$$\gamma = \sqrt{\frac{K_\varnothing}{K_\theta}} \tag{17.2}$$

where K_θ is the pressure loss coefficient at blade angle θ,
$K_\varnothing$ is the pressure loss coefficient at blade angle $\varnothing$.

INSTALLED CHARACTERISTIC

The installed characteristic γ' is the relationship between the stroke of the FCD and the flow rate through the system relative to the flow rate with the FCD when fully open. The general case is where the total pressure available in the system varies with the flow rate. An example of this is when the control device is used to *throttle* the total flow rate of a pump or fan. If the prime mover characteristic is defined by Eq. (15.7) in Chapter 5, then:

$$\gamma' = \sqrt{\frac{r_\varnothing / N + B'}{r_\varnothing \left(\frac{1}{N} + \frac{1}{\gamma^2}\right) + B'}} \tag{17.3}$$

For a system with a constant pressure drop $B' = 0$, the installed characteristic becomes:

$$\gamma' = \frac{\gamma}{\sqrt{[N + \gamma^2(1 - N)]}} \tag{17.4}$$

Thus, for a given design (e.g., fan/system) and inherent characteristic, there is a set of installed characteristic curves for the FCD, each member of which is determined by the authority N. A typical set of curves is illustrated in Fig. 17.1.[1]

---

[1] The derivations of the Eqns. (17.2)–(17.4), and other equations later in the chapter are given in Appendix 1.

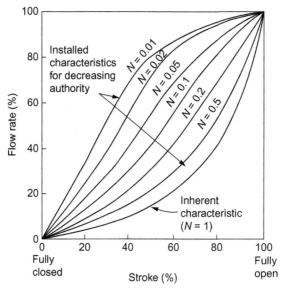

Fig. 17.1 Inherent and installed characteristics of a modulating fluid flow control device.

DAMPERS

If a system is to operate satisfactorily, the design and selection of a damper should be treated with appropriate importance relative to the other components within the system. Unfortunately, this has often been neglected in the past, as few manufacturers publish the necessary data to allow a selection to be made, and the damper is often sized to fit the ductwork, which has been sized to accommodate other plant items.

Damper Pressure Loss Coefficients

Control dampers in most common use for ducted air systems are normally made for rectangular ducts. These consist of a number of blades each with a central spindle linked together to allow all the blades to move together varying the angle of inclination θ to the axis of the duct, as shown in Fig. 17.2. With two or more blades, the spindles are linked to give either parallel or opposed movement, over a wide range of blade angles, from a start angle of φ degrees relative to the duct axis.

Commercial dampers have a variety of cross-sectional blade shapes including flat plates with crimped edges, aerofoil, and wedge. The leading and trailing edges of the blades often incorporate material to ensure a reasonable degree of air tightness when in the closed position. For relatively

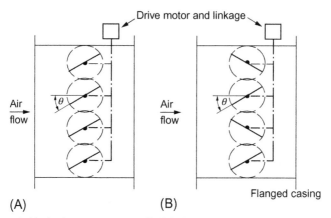

Fig. 17.2 Multiblade dampers. (A) Parallel blades and (B) opposed blades.

small circular and rectangular ducts, a single-leaf damper is sometimes referred to as a butterfly damper.

It has also been shown experimentally [1] that the damper loss coefficients $K_{d\theta}$ are defined by the empirical relationship:

$$\log_e K_{d\theta} = a + b\theta \tag{17.5}$$

Generally, in the blade angle ranges:

Single and opposed blades: 10 degrees $\leq \theta \leq$ 60 degrees
Parallel blades: 10 degrees $\leq \theta \leq$ 70 degrees

This relationship is illustrated in Fig. 17.3. The value of the experimental constants a and b relates to parameters such as blade size and shape, projections on the blade, relationship between the duct wall, and blade configuration.

Since the inherent characteristic for a damper may be obtained from pressure loss coefficients (Eq. 17.2), it follows from Eq. (17.5) that the inherent damper characteristic becomes an exponential decay, i.e.,

$$\gamma = \sqrt{\frac{K_{d\varnothing}}{K_{d\theta}}} = \sqrt{\frac{e^{(a+b\varnothing)}}{e^{(a+b\theta)}}}$$

$$\therefore \gamma = e^{(\varnothing - \theta)b/2} \tag{17.6}$$

For Eq. (17.6) to be applied for blade angles <10 degrees, then the blade and/or duct wall projections should be sized for:

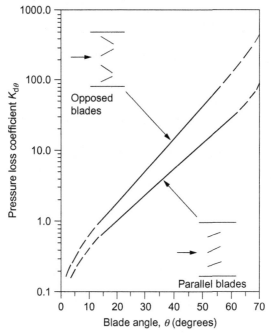

Fig. 17.3 Damper pressure loss coefficients.

$$K_{d\varnothing} = e^a \qquad (17.7)$$

where $\varnothing$ is the zero degrees.

Typical values of the constants a and b for crimped blade dampers are given in Table 17.1 below.

Damper Size and Selection

It is generally accepted that a linear installed characteristic is desirable for flow regulation in air handling systems. To obtain a *near*-linear installed

Table 17.1 Typical values of empirical constants a and b (Eq. 17.5) for crimped blade dampers, together with damper sizing constants, G_d

Type	Number of blades	Constants a	b	Damper sizing constant G_d $\varnothing=0°$	$\varnothing=10°$
Single	1	−1.5	0.105	0.50	0.39
Opposed	≥2	−1.5	0.105	0.50	0.39
Parallel	2	−1.5	0.088	0.75	0.61
Parallel	>3	−1.5	0.080	0.91	0.77

characteristic where the system pressure drop varies with flow rate, an iterative solution of Eq. (17.3) will normally be required. However, the particular but typical case is where a *part* of the system (in which the damper controls the flow rate) operates at a constant pressure drop. Examples of this are the three dampers $D_1, D_2,$ and D_3 in Fig. 17.4 where the pressures at nodal points **X** and **Y** can be assumed to be constant for a CAV system.

Referring to Fig. 17.5, if an installed characteristic is required, which fits closest to the linear relationship **AB**, the authority can be calculated, which minimizes the shaded areas, and a close approximation for this would be for $\gamma' = 0.5$ at a midrange blade angle of $(90 - \varnothing)/2$. Therefore, rearranging Eq. (17.4) the authority is given by

$$N = \frac{3\gamma^2}{1 - \gamma^2} \qquad (17.8)$$

It can then be shown that

$$\dot{V} = A_{\mathrm{d}} G_{\mathrm{d}} \sqrt{\Delta p_{\mathrm{s}}/\rho} \qquad (17.9)$$

$$G_{\mathrm{d}} = \sqrt{\frac{2N}{K_{d\varnothing}(1 - N)}} \qquad (17.10)$$

where G_{d} is the constant for a particular damper design and A_{d} is the damper cross-sectional area.

Note: Eq. (17.9) is similar to the valve sizing equation (Eq. 17.12.)

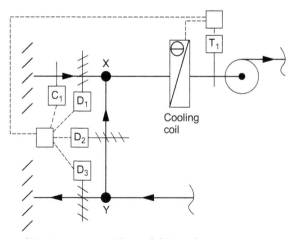

Fig. 17.4 Air conditioning system with modulating dampers.

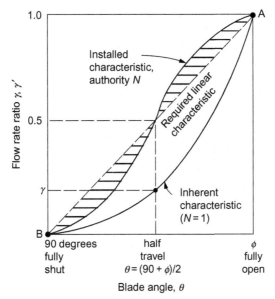

Fig. 17.5 Sizing a damper to give a near-linear installed characteristic.

The loss coefficient $K_{d\varnothing}$ is determined from Eq. (17.5). The authority, N, is obtained from Eq. (17.8) in which the inherent characteristic is defined by Eq. (17.6) at the midrange angle of $(90 - \varnothing)/2$.

Two examples of the use of these equations are given below.

Example 17.2

For an opposed blade damper using empirical values of $a = -1.5$ and $b = 0.105$, determine the value of the damper sizing constant, G_d, for a start angle of zero degrees.

Solution

For a start angle of *zero* degrees, the damper should be designed for the loss coefficient $K_{d\varnothing}$ obtained from Eq. (17.7):

$$K_{d\varnothing} = e^a$$
$$= e^{-1.5} = 0.223$$

At $\theta = 45$ degrees, the value at $\gamma' = 0.5$.
With $\varnothing = 0$ and using Eq. (17.7):

$$\gamma = e^{(\varnothing - \theta)b/2}$$
$$= e^{(0 - 50)0.105/2}$$
$$= 0.0942$$

Using Eq. (17.8):

$$N = \frac{3\gamma^2}{1 - \gamma^2}$$

$$N = \frac{3 \times 0.0942^2}{1 - 0.0942^2}$$

From Eq. (17.10), the damper sizing constant is therefore obtained:

$$G = \sqrt{\frac{2N}{K_{d\varnothing} (1 - N)}}$$

$$G = \sqrt{\frac{2 \times 0.0942}{0.223 (1 - 0.0269)}} = 0.498$$

Example 17.3

For a parallel crimped blade damper using typical empirical values $a = 1.5$ and $b = 0.08$, determine the value of the damper sizing constant G with a start angle $\varnothing$ of 10 degrees.

Solution

The loss coefficient $K_{d\varphi}$ is given by Eq. (17.5):

$$\log_e K_{d\varnothing} = a + b\varnothing$$

$$= -1.5 + 0.08 \times 10$$

$$\therefore K_{d\varnothing} = 0.0497$$

At $\theta = 50$ degrees, the value at $\gamma' = 0.5$.
Using Eq. (17.6):

$$\gamma = e^{(\varnothing - \theta)b/2}$$

$$= e^{(10 - 50)0.08/2}$$

$$= 0.202$$

Using Eq. (17.8):

$$N = \frac{3\gamma^2}{1 - \gamma^2}$$

$$= \frac{3 \times 0.202^2}{1 - 0.202^2} = 0.1275$$

Using Eq. (17.10):

$$G = \sqrt{\frac{2N}{K_{d\varnothing} (1 - N)}}$$

$$G = \sqrt{\frac{2 \times 0.1275}{0.497\,(1 - 1275)}} = 0.767$$

Typical values of G are given in Table 17.1.

Use of Damper Sizing Equation

An example of sizing dampers for a 'free-cooling' recirculation system (see Chapter 6) is given below.

Example 17.4

For the system shown in Fig. 17.4, determine the sizes of the three dampers if the design flow rates are 2.0 m/s for the supply and 1.8 m/s for exhaust and recirculation. The pressure drops have been calculated to give the following internal duct air pressures relative to atmospheric pressure:

At nodal point **X**   −30 Pa
At nodal point **Y**   +40 Pa

Solution

For a damper start angle of zero degrees, an air density of 1.2 kg/m³, and using Eq. (17.9), the calculation is set out in the following table (values of G_d taken from Table 17.1):

Parameter	Opposed blade damper			Parallel blade damper		
	D_1	D_2	D_3	D_1	D_2	D_3
Damper sizing constant, G_d		0.50			0.91	
Flow rate, V (m³/s)	2.0	1.8	1.8	2.0	1.8	1.8
System pressure drop (Pa)	30	70	40	30	70	40
Damper cross section, A_d (m²)	0.80	0.47	0.63	0.44	0.26	0.34

Choice of Parallel or Opposed Blade Dampers

Example 17.4 shows that, for a given design system flow rate and pressure drop, an opposed blade damper will be approximately twice the size of a parallel blade damper. Having considered this, the choice of damper will then depend on a number of other factors, including the following:

- Convenience of making the duct connections—the ductwork sized for other plant items may more easily accommodate one damper rather than another.
- The costs of the alternative arrangements.
- The importance of the length of duct required for full static pressure regain, e.g., pressure controllers might be influenced adversely in the low-pressure region immediately downstream of the damper; full pressure recovery for an opposed blade damper occurs between three and six equivalent duct diameters downstream of the damper, 10 diameters for a parallel blade damper [2].
- The ability of the dampers to mix two airstreams at two temperature levels and to prevent stratification in the total air volume flow before the next section of the system. This is particularly the case where the dampers are arranged to discharge directly into a mixing chamber.

With regard to this last point, the alternative arrangements to achieve mixing are shown in Fig. 17.6. For throttling applications, provided they are sized at an appropriate authority, opposed and parallel blade dampers will work equally well. Note that for plenum discharge, the empirical values of a and b are likely to be rather higher than those given in Table 17.1.

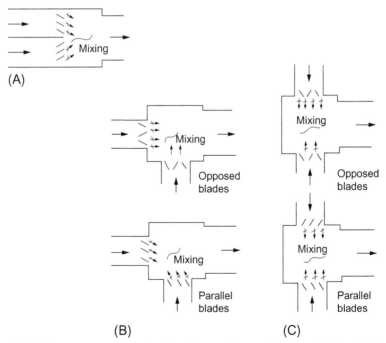

Fig. 17.6 Mixing arrangements for inlet ducts with two dampers. (A) Parallel flows, (B) right angle flows, and (C) opposite flows.

Leakage

When considering liquid control valves, the ratio of maximum flow to minimum controllable flow at constant pressure drop across the valve is known as *rangeability*. When the valve is installed in the system, its authority affects this minimum flow, and *turndown is* the term used to define the ratio of maximum flow to minimum controllable flow with a constant pressure drop across the circuit. These terms have not yet been accepted as part of damper performance. It is more usual to refer to the flow through the damper in the fully closed position as *air leakage* for both the inherent and installed characteristics. For efficient plant operation, leakage should be minimized wherever complete shut off is required; e.g., this will normally be the case for the recirculation damper D_2 in Fig. 17.4.

CONTROL VALVES

John F. Missenden

Emeritus Professor of Engineering Systems, London South Bank University
Regulation of cooling and heating systems in building services takes place mainly at the heat exchanger, through the adjustment of fluid flow rates and temperatures. The modulation of a fluid supply temperature can be accomplished by on/off control of boilers and chillers resulting in a cyclical or 'saw-tooth' temperature pattern. The smooth variation in temperature usually required is achieved more effectively by mixing streams of fluids at different temperatures. A valve of the three-port mixing type is needed for this purpose just as a two-port control valve is, on first sight, required for flow rate modulation. In practise, the main method used for the control of thermal output is, in both cases, the variation of a valve opening.

Where the output of a heat exchanger is regulated by varying the flow rate, the primary medium inlet temperature to the heat exchanger remains constant. Therefore, the output of heat only falls off because of the reduction in heat transfer coefficients, which in turn depend on the reducing flow rate. This is offset, somewhat, by the increase in temperature drop consequent upon the longer primary medium residence time in the heat exchanger. As a result, flow rate modulation is of a very nonlinear form, with typically 80% of flow reduction needed for a 40%–50% heat output drop. In this case, controllers can find difficulty in producing accurate and stable outputs.

It follows that there are several conflicting requirements in engineering the application of control valves. On the one hand, close, linear, stable regulation of output is needed, but on the other hand, liquid circulation requires system balance, freedom from sediment, minimizing pump and ancillaries sizes, and power use. Valve types, characteristics, sizing, and circuit design all contribute to obtaining an optimum solution. The two main types of control valve are flow restrictors and flow mixers.

Flow Restrictors

A flow restrictor is a two-port valve that reduces the flow as the valve stem is lowered and the plug pushed onto the valve seat. The disadvantage of this type of valve is that flow may be entirely obstructed. This in turn leads to the pump having to operate against closed valve conditions and to boilers and water chillers operating with insufficient flows. Scaling and deposition of solids tend to occur, in the worst possible place, i.e., on the exchanger heat transfer surfaces, leading to permanent loss of output and eventual blockage of the small tubes. Maximum pressure drops against which the valve may close are also limited, particularly where system pressure may rise steeply as the pump operates at a lower flow rate. These objections to two-port valves do not apply in the case of steam flow regulation. In this case, the valve may be used directly to govern a heat exchanger output required for air or domestic hot water heating. The relatively high velocity and modest pressure of process steam obviate the difficulties otherwise experienced.

Flow Mixers

A flow mixing valve is one in which the stem position is altered, the opening of one inlet port increases whilst the flow area of the other decreases. The blend of fluid varies progressively from one inlet condition toward the other; hence, this is known as a *three-port changeover* valve. The temperature t_m of the mixed outlet fluid is given by:

$$t_m = \frac{\dot{m}_1 t_1 + \dot{m}_2 t_2}{\dot{m}_1 + \dot{m}_2} \qquad (17.11)$$

where $\dot{m}_1$ and $\dot{m}_2$ are the inlet mass flow rates and t_1 and t_2 are the inlet temperatures, respectively.

In general, mixing valves should be designed to deliver a constant outlet mass flow rate irrespective of stem position. This allows the outlet temperature to be the principal variable, independent of mass flow effects, a particular requirement for good heat exchanger control.

The near-constant mass flow rate from a mixing valve is an advantage in the part of the circuit following the valve. However, both inlet flows will vary, and provision may have to be made in a pumped primary circuit to allow an alternative fluid return path when all valves are closed to flow from the main. An example of this is the pressure relief valve that allows a primary main to discharge direct to the return main if flow rate becomes insufficient and the pressure difference between flow and return mains becomes excessive.

PIPE CIRCUITS

Two main types of circuit exist to which three-port mixing valves may be applied. Each produces a distinctly different effect that must be understood if the piping/valves are to be designed and sized correctly.

Mixing

If the control valve is situated in the heat exchanger supply branch to mix primary fluid with return fluid, as shown in Fig. 17.7, then the circuit is said to be of the mixing type. In order for the return fluid to mix, pressurization is required by means of subcircuit pumping. A balancing value is also required in the return branch to equalize pipe pressure losses around the alternative circuits. Frequently, a balance pipe is installed to avoid interaction between primary and subcircuit pumps.

The blending of the two flows maintains a constant mass flow rate to the heat emitters whilst allowing linear variation in temperature. With heat transfer equipment, there is a near-linear relationship between output and flow water temperature—a satisfactory characteristic. Conveniently, linear

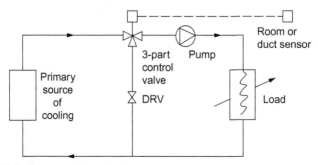

Fig. 17.7 Chilled water mixing circuit.

valve and actuator characteristics may therefore be employed, leading to good control though more expensive than the diverting circuits described below. Acceptable authorities are held to lie in the range $0.5 < N < 0.7$.

Diverting

To avoid the need and expense of subcircuit pumping, the alternative position for a mixing valve may be chosen, i.e., in return, as shown in Fig. 17.8. Here, return fluid from the heat emitters is mixed with bypass fluid to give a blended return to the primary system. Bypass balance must be achieved by a DRV to maintain constant total flow at all control valve positions. This is a simpler and more economical arrangement than the mixing circuit.

Because of the mixed return flow, the change in the subcircuit is now made to the flow rate rather than the temperature flow. In this case, the output of the heat exchange equipment is not linear to flow rate. As the flow rate reduces, the fluid's time inside the exchanger increases, and so the contact factor (or effectiveness) and hence temperature change increase. This nonlinearity has to be taken into account when selecting the appropriate valve/actuator characteristics. Acceptable authorities for these applications are in the range $0.3 < N < 0.5$. The diverting circuit is more demanding in terms of technical design but is cheaper in capital and running costs.

VALVE CHARACTERISTICS

A valve test consists of measuring the pressure drop through a valve for a range of flow rates at different positions of the valve stem (or percentage

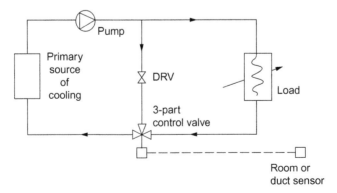

Fig. 17.8 Chilled water diverting circuit.

opening), for authorities of unity. When these points are plotted, the valve inherent characteristic is obtained. Three broad types of characteristic are found, each of which has a particular application; these are described below.

Linear Valves

In practise, the installed characteristic of a linear valve may be only approximately linear at normal authorities. As the authority falls below unity, every valve becomes more *quick opening*, and the inherent (or test) characteristic should therefore be only moderately parabolic in shape. Not surprisingly, the shape of the valve plug/seat combination mirrors that of the flow characteristic. Valve plugs of this type are shown in Fig. 17.9A and B, and the installed characteristics for various authorities are shown in Fig. 17.1.

The applications of this type of valve plug are in three-port valves used in mixing circuits, where linear changes in blending ratio produce linear variations in, say, a heat exchanger output. Two-port steam control valves should also have a generally linear behaviour, as flow rate is directly proportional to output.

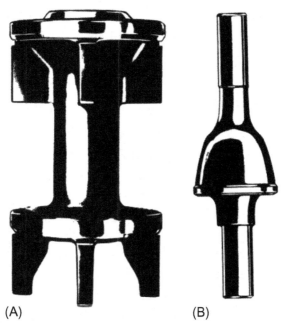

(A) (B)

Fig. 17.9 Plugs for linear valves. (A) Double port, (B) single port.

Quick-opening Valves

A quick-opening valve, also known as a *bevel disk* type of valve, has the characteristic of giving near maximum flow for a relatively low stem lift. The gate valve is usually of this type and, combined with its poor authority, is difficult to use for manual flow regulation. An example of the type of plug is shown in Fig. 17.10, and typical characteristics are shown in Fig. 17.11.

The lift needs to only be about 25% of the maximum, usually of the same order as the diameter, to allow full flow. This type of characteristic is ideally suited to the solenoid operated on/off type of valve. The sharpness of operation may bring some risk of water hammer where there are long pipe runs and high static pressures.

Equal Percentage Valves

As the contact factor or effectiveness of a heat exchanger rises as the flow rate falls, a valve is needed that mirrors and therefore reverses this exponential effect. Because heat output falls only slowly as feed is reduced, a valve characteristic that is opposite to this is called for, i.e., a valve that increases the flow rate slowly as it opens, with perhaps 15%–20% of full flow at the midstem position. An *equal percentage* valve is designed to give this effect, and the characteristics are shown for various authorities in Fig. 17.12.

Fig. 17.10 Double port plug for a quick-opening valve.

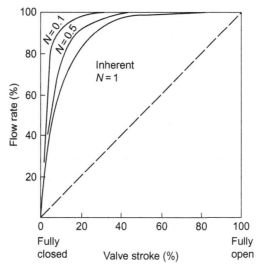

Fig. 17.11 Typical quick-opening valve installed characteristics.

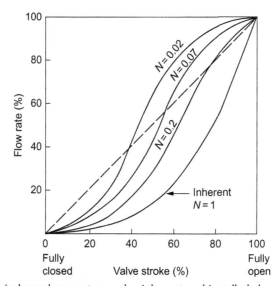

Fig. 17.12 Typical equal percentage valve inherent and installed characteristics.

The application most desirable for this type is in the mixing valve used in a diverting circuit. In order to retain system balance, the combined (or mixed) total flow through such a valve must stay reasonably constant for any stem position. For this reason, only the valve plug facing the heat exchanger should be of equal percentage type.

VALVE SIZING

As pointed out earlier in the chapter, the sizing of a valve is similar to the sizing of a damper. The valve sizing equation is:

$$Q = K_v \sqrt{\Delta p_s} \qquad (17.12)$$

where Q is the flow rate, K_v is the valve sizing constant, and Δp_s is the system pressure drop.

Notes:

- The system pressure drop Δp_s, is the pressure drop in that *part of the system* in which the valve controls the flow rate, not necessarily the whole system.
- Equation (17.12) is for use with water; since water density can be considered constant at 1.0 kg/m^3 (at 4°C), it is incorporated in the valve sizing constant.
- If other liquids are used, the liquid density should be incorporated in Eq. (17.12) (see Eq. 17.10).
- The symbol C_v is often used as the valve sizing constant, instead of K_v.
- Manufacturers provide the valve sizing constants.

SYMBOLS

A', B'	constants of fan characteristic
A	representative area of FCD
A_d	area of damper cross section
a	intercept of Eq. (17.5)
b	slope of Eq. (17.5)
CAv	constant air volume
D	plug diameter
FCD	flow control device
G_d	damper sizing constant (defined by Eq. 17.9)
K	pressure loss coefficient (defined by Eq. (13.12))
K_v	valve sizing constant
L	stem lift
$\dot{m}$	mass flow rate
N	authority
r	resistance
t	temperature
VAV	variable air volume
$\dot{V}$	volume flow rate
$\bar{v}$	mean velocity
γ	inherent characteristic
γ'	installed characteristic
Δp	pressure drop

ρ air density

θ valve (stem) position or damper blade angle

$\emptyset$ valve (stem) fully open or damper start angle

SUBSCRIPTS

d   damper

s   system

θ position of damper blade angle or valve stroke

$\emptyset$ initial position of start angle of damper blades or valve stroke (valve or damper fully open)

REFERENCES

[1] R.C. Legg, Characteristics of single and multiblade dampers for ducted air system, Build. Serv. Eng. Res. Technol. 7 (1987). p. 129.

[2] R.C. Legg, Multiblade dampers for ducted air systems, in: IMechE Conference: Installation Effects in Ducted Fan Systems, 1984, p. 67.

CHAPTER 18

Energy Consumption

The evaluation of the annual energy consumption (and energy costs) of air-conditioning systems is required for the following reasons:

- To allow a comparison of systems on a total cost basis;
- To optimize the design and control of the selected system to achieve minimum energy consumption;
- To inform the client of the expected costs for budgeting purposes;
- To provide a basis for the energy performance of the installed system during its working life.

The method of calculating the average annual energy consumption for heating systems using the concept of degree days is well established [1]. There is no such commonly accepted simplified method for air conditioning systems as there can be a poor relationship between heating and cooling degree days and energy consumed due to overriding factors including fresh air requirements and humidity control. Computer models are commonly available that provide dynamic thermal simulations of building; systems using historical weather data can be used to assess the energy consumption. However, a simple approach, known as the BIN method, provides a useful method of energy assessment that may be particularly useful when comparing systems in the early stages of the design process or where more complex software systems are not available.

THE BIN METHOD OF CALCULATING ANNUAL ENERGY CONSUMPTION

The BIN method is a technique to estimate annual energy consumption from which cost comparisons between alternative designs can be made. The technique is similar to that given in the ASHRAE Handbook [2], which uses a base of dry-bulb temperature (5°F/2.8°C intervals). Provided the relevant data are available, the approach described below offers greater flexibility in the choice of outdoor conditions. It is often practically more useful than the simpler degree-day methods as is allows estimations of energy use where the cooling or heating loads vary nonlinearly with outdoor

Air Conditioning System Design
http://dx.doi.org/10.1016/B978-0-08-101123-2.00018-2

temperature. It can also provide a swifter and more transparent visualization of comparative energy use in systems than with more complex software simulations, allowing early assessment of design options.

The predicted annual energy usage for a complete air conditioning system is obtained through the summation of the energy used by each individual energy-using plant item within the system. These may include the following:

- Heaters and boiler plant;
- Coolers and refrigerating plant;
- Steam humidifiers;
- Fans for central plant supply and extract;
- Fans for room air conditioning units;
- Fans for cooling towers;
- Pumps for heating, cooling, and humidification water systems;
- Compressors for unitary systems.

Some of these plant items will run at constant load in which case the annual consumption is simply the annual hours of operation multiplied by the installed load. Where plant load varies because of variations in outdoor conditions and internal heat gains, it will be necessary to analyse the load variations according to their annual frequency of occurrence.

The method is as follows:

a) Establish the relationship between energy demand and an appropriate climate factor;

b) Integrate the variations in energy demand with the annual frequency distribution of the climate factor (such as the frequency data described in Chapter 4, these data are used in the examples later in the chapter).

The general principles of the BIN method are outlined using Fig. 18.1. The load on the plant item being considered is plotted against the outdoor climate factor for which there is a known frequency of occurrence. Since the load is increasing with outdoor conditions, this diagram represents the load handled by a cooler battery in an air conditioning system. The main features of this diagram are the following:

- The load profile is shown on the upper chart as **ABCDE**.
- The plant is switched on at **A**.
- The load between points **A** and B is constant and is termed a *base load*. A base load may be due to miscellaneous heat gains and losses to/from pipework. The frequency of such occasions is obtained by summing

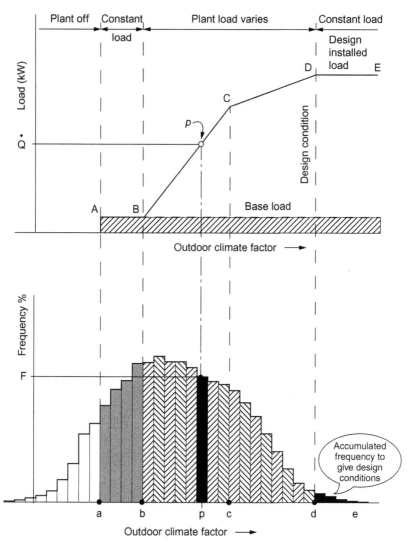

Fig. 18.1 Plant load vs outdoor climate factor with known frequency of occurrence—BIN method.

each of the frequency of occurrences between points **A** and **B**—this is identified on the lower chart between points **a** and **b**.

• The load varies between **B** and **C** in the frequency range **be**. The mode of operation of the plant changes at **C**, and the load varies between **CD** in the frequency range **cd**.

- For any load, the expected frequency may be obtained; e.g., at point **P**, load **Q** occurs with frequency **f**.
- The load **DE** is constant at the design or installed load (i.e., the plant is unable to increase the load above this). This occurs for a total accumulated frequency in the range **de**.
- The choice of climate factor will depend on the preferred operating characteristics of the system.
- A *bin*, used in the BIN method for calculating energy consumption, is any one of the frequency columns of the lower chart; any plant load, part-load efficiency, and heat gain/loss associated with that particular range of the climate factor are *placed in the bin*.

Where both the load and frequency of occurrence of the load are varying, it is necessary to accumulate the sum of their product. The average annual energy consumption is then given by the following:

For cooling loads,

$$E_c = \frac{H}{100} \sum \left(\frac{fQ}{COP} \right) \qquad (18.1)$$

For heating loads,

$$E_c = \frac{H}{100} \sum \left(\frac{fQ}{\eta} \right) \qquad (18.2)$$

where COP is the cooling system part-load coefficient of performance, η is the heating system part-load efficiency and H is the operating time of plant in hours

From this calculation, the annual energy consumption is in units kWh; the annual energy costs may be obtained from the unit costs of the fuel.

This method is particularly useful for the analysis of the cost benefit of additional energy saving equipment in a system. In this case, only the analysis of the *difference* in energy costs between the alternative schemes would be required.

For the calculations, a range (or interval) of the climate factor is chosen appropriate to the system. Too small a range leads to a lengthy calculation; too large a range leads to reduced accuracy and flexibility.

A number of examples now follow that illustrate the general approach and method of calculation.

Example 18.1

The air conditioning system (system 1) shown in Fig. 18.2 provides a constant ratio of outdoor to recirculated air quantities, and the off-coil condition **B** is maintained by the dew-point sensor T_{dp}. Calculate the energy consumption for the cooler for the following criteria:

Mass air flow rate (assumed constant)	2.5 kg/s
Air enthalpies	
Recirculation, h_i	42 kJ/kg$_{da}$
Off-coil, h_b	30 kJ/kg$_{da}$
Outdoor air design condition	54 kJ/kg$_{da}$
Outdoor air mass flow rate fraction (x)	0.25 kg/s
Coil contact factor	1.0
Climate data for Heathrow (Table 4.4)	
Hours of plant operation per annum	8760
COP (assumed constant)	3.8

The operation of this system is explained in Chapter 6.
Assume no miscellaneous heat gains or losses from fans, ductwork, or pipes.

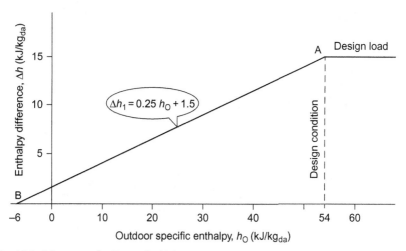

Fig. 18.2 AC system for Example 18.1.

Solution

Plot typical psychrometric processes—Fig. 18.3. It is seen from this that in order to maintain the space conditions, the cooler will be required to operate throughout the year down to low outdoor air enthalpies.

The year-round cooling load is given by the following:

$$Q = \dot{m}_a \left(h_M - h_B \right)$$

Since the mass flow rate, $\dot{m}_a$, remains constant, this can be omitted temporarily from the calculations until the numerical integration has been completed. Therefore, the load can be expressed in terms of an enthalpy difference in the following:

$$\Delta h_1 = h_M - h_B$$

$$\Delta h_1 = (1-x)h_B + x h_o - h_B$$

$$= (1-0.25)\,42 + 0.25 h_o - 30$$

$$\Delta h_1 = 0.25 h_o + 1.5 \tag{18.3}$$

Using Eq. (18.3), the design load is given by the following:

$$\Delta h_d = 0.25 \times 54 + 1.5 = 15\,\text{kJ/kg}_{da}$$

The load profile for the operation of this system is given in Fig. 18.4.

The numerical integration is completed in Table 18.1, with the frequency of occurrence of hourly values of the outdoor air climate factor, grouped in convenient ranges of specific enthalpy from Table 4.3 in Chapter 4. The notes column allows the engineer to keep track of the operating conditions of the plant.

The accumulated sum of the product is $(f_3 \Delta h_1) = 812.5$

Using Eq. (18.1), the annual energy consumption is given by the following:

$$E_c = \frac{H}{100} \sum \left(\frac{f_3 Q}{\text{COP}} \right) 100$$

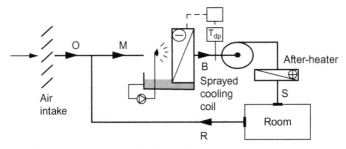

Fig. 18.3 Psychrometric processes for Example 18.1.

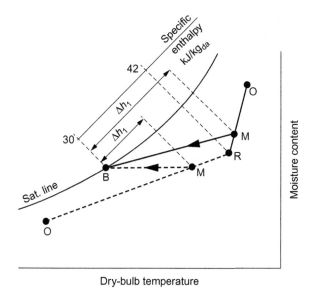

Fig. 18.4 Load profile—Example 18.1.

Table 18.1 Calculation of average hourly energy demand of cooler operating in system 1, Example 18.1

Range of outdoor air specific enthalpies (kJ/kg$_{da}$)	Midpoint of range h_o (kJ/kg$_{da}$)	Frequency f_3 (%)	Δh_1	$f_3 \Delta h_1$	Notes
<2.0	−3	0.32	0.75	0.2	Use Eq. (18.3)
2–5.9	4	1.38	2.5	3.5	$\Delta h_1 = 0.25 h_o + 1.5$
6–9.9	8	4.65	3.5	16.3	
10–13.9	12	7.78	4.5	35.0	
14–17.9	16	10.41	5.5	57.3	
18–21.9	20	12.38	6.5	80.5	
22–25.9	24	12.68	7.5	95.1	
26–29.9	28	12.19	8.5	103.6	
30–33.9	32	11.12	9.5	105.6	
34–37.9	36	10.42	10.5	109.4	
38–41.9	40	8.12	11.5	93.4	
42–45.9	44	5.15	12.5	64.4	
46–49.9	48	2.27	13.5	30.6	
50–53.9	52	0.84	14.5	12.2	
>54.0	–	0.36	15.0	5.4	Design load
Total	$\sum (f_3 \Delta h_1)$			812.5	

Since $\dot{m}_a$ and COP are constant:

$$\therefore E_c = \frac{\dot{m}_a \, H}{\text{COP} \, 100} \sum f_3 \, \Delta h_1$$

$$= \frac{2.5 \times 8760}{3.8 \times 100} \times 812.5 = 46,826 \, \text{kWh/annum} \, (169 \, \text{GJ})$$

Example 18.2

In the recirculation air conditioning system (system 2) shown in Fig. 18.5, the cooler and the dampers $D_1, D_2,$ and D_3 are operated sequentially through the sensor T_{dp} to maintain the off-coil condition **B**. The damper changeover is achieved through C_1 with a set point at the room condition of 42 kJ/kg. Calculate the annual energy consumption of the cooler for the data given in Example 18.1. (The operation of this system is described in Chapter 6.)

Solution

Plot typical psychrometric processes, as in Fig. 18.6:

Summer operation—as system 1

Midyear operation—100% outdoor air

Winter operation—dampers modulate, cooler not in operation.

With the outdoor air condition above the set point of C_1, the cooling load is the same as for system 1 in Example 18.2, i.e., use Eq. (18.3). When the

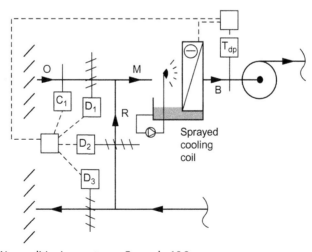

Fig. 18.5 Air conditioning system—Example 18.2.

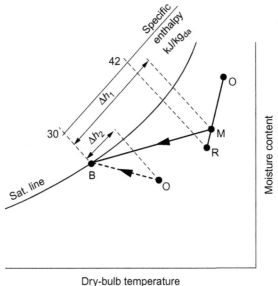

Fig. 18.6 Psychrometric processes for Example 18.2.

outdoor air enthalpy is between the set points of T_1 and C_1, the system operates on 100% outdoor air $(x=1.0)$, and the cooling load becomes the following:

$$Q = \dot{m}_a(h_O - h_B)$$

Therefore, as the mass flow rate is constant:

$$\Delta h_2 = h_O - 30 \qquad (18.4)$$

The load profile for the operation of this system is shown in Fig. 18.7.
The numerical integration is completed in Table 18.2.
The accumulated sum of the product $(f_3 \Delta h_2) = 278.5$.
Using Eq. (18.1), the annual energy use is obtained:

$$E_c = \frac{H}{100} \sum \left(\frac{fQ}{\text{COP}} \right)$$

Since the mass flow rate $\dot{m}$ is constant:

$$
\begin{aligned}
E_c &= \frac{\dot{m}H}{\text{COP} \times 100} \sum (f \Delta h_2) \\
&= \frac{2.5 \times 8760}{3.8 \times 100} \times 278.5 = 16{,}050 \, \text{kWh/annum} \, (57.8 \, \text{GJ})
\end{aligned}
$$

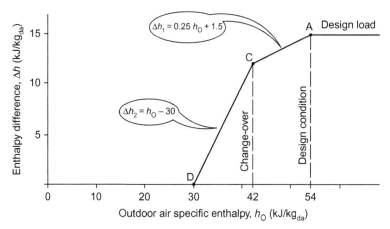

Fig. 18.7 Load profile—Example 18.7.

Table 18.2 Calculation of average hourly energy demand of cooler operating in system 2 (Example 18.2)

Range of outdoor air specific enthalpies (kJ/kg$_{da}$)	Midpoint of range h_o (kJ/kg$_{da}$)	Frequency f_3(%)	Δh_1	$f_3\,\Delta h$	Notes
<30	–	–	–	–	Cooler off
30–33.9	32	11.12	2.0	22.2	Use Eq. (18.4)
34–37.9	36	10.42	6.0	62.5	$\Delta h_2 = h_O - 30$
38–41.9	40	8.12	10.0	81.2	
42–45.9	44	5.15	12.5	64.4	Use Eq. (18.3)
46–49.9	48	2.27	13.5	30.6	$\Delta h_1 = 0.25 h_o + 1.5$
50–53.9	52	0.84	14.5	12.2	
>54.0	–	0.36	15.0	5.4	Design load
Total	$\sum (f_3\,\Delta h)$			278.5	

The two systems in Examples 18.1 and 18.2 are alternative methods of treating the recirculated air, the inclusion of modulating dampers making system 2 more energy-efficient than system 1. The difference between the energy consumption between the two systems is then given by the following:

$$46,826 - 16,050 = 30,776 \, \text{kWh/annum} \, (111 \, \text{GJ})$$

The theoretical percentage energy saving, expressed as a percentage, is in the following:

$$\frac{30,776}{46.826} \times 100 = 66\%$$

The difference in annual energy consumption can be costed as part of a cost benefit calculation for the additional dampers, ductwork, and controls.

Short Route to Determine Energy Consumption Difference

If only the *difference* between the systems is required, this can be obtained by a shorter route as shown in the following example.

Example 18.3

Determine the difference in energy consumption between systems 1 and 2 of Examples 18.1 and 18.2

Solution

The load profiles for systems 1 and 2 are shown in Figs 18.4 and 18.7, respectively. If these two diagrams are superimposed, Fig. 18.8 is obtained; the difference in energy consumption is then represented by the area **BDC**. The intervals of enthalpy difference of this area are then integrated with the frequency of specific enthalpy in convenient ranges.

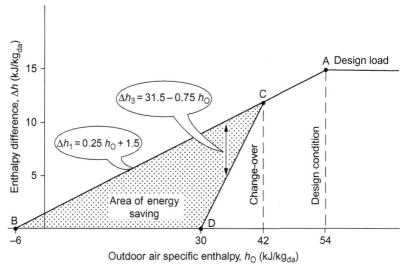

Fig. 18.8 Load profile—Example 18.3.

Table 18.3 Calculation of the difference in average hourly energy demand of cooler operation, system 1 compared with system 2 (Example 18.3)

Range of outdoor air specific enthalpies (kJ/kg$_{da}$)	Midpoint of range h_o(kJ/kg$_{da}$)	Frequency f_3(%)	Δh	$f_3 \, \Delta h$	Notes
<2.0	−3	0.32	0.75	0.2	Use Eq. (18.3)
2–5.9	4	1.38	2.5	3.5	$\Delta h_1 = 0.25\,h_o + 1.5$
6–9.9	8	4.65	3.5	16.3	
10–13.9	12	7.78	4.5	35.0	
14–17.9	16	10.41	5.5	57.3	
18–21.9	20	12.38	6.5	80.5	
22–25.9	24	12.68	7.5	95.1	
26–29.9	28	12.19	8.5	103.6	
30–33.9	32	11.12	7.5	83.4	Use Eq. (18.5)
34–37.9	36	10.42	4.5	46.9	$\Delta h_3 = 31.5 - 0.75\,h_o$
38–41.9	40	8.12	1.5	12.2	
Total	$\sum (f_3 \, \Delta h)$			534.0	

For $\Delta h < 30$, Eq. (18.3) applies.
In the enthalpy range of 30–42 kJ/kg$_{da}$, the difference is given by the following:

$$\Delta h_3 = \Delta h_1 - \Delta h_2$$

$$= (0.25\,h_O + 1.5) - (h_O - 30) \tag{18.5}$$

$$\Delta h_3 = 31.5 - 0.75\,h_O$$

The numerical integration is completed in Table 18.3, and the accumulated sum of the product is $(f_3 \Delta h) = 534.0$.
Using Eq. (18.1), the annual energy saving, from system 1 to system 2, is obtained:

$$E_c = \frac{H}{100} \sum \left(\frac{fQ}{COP} \right)$$

$$E_c = \frac{\dot{m}_a \, H}{COP \times 100} \sum (f \Delta h)$$

$$= \frac{2.5 \times 8760}{3.8 \times 100} \times 534 = 30{,}776\,\text{kWh/annum}$$

which is (unsurprisingly!) the same as the energy saving calculated previously.

Energy Saving of a Heat Recovery Unit

The following example considers a system heating requirement including an air-to-air heat recovery unit (HRU); a heater is required after the HRU as the output of this unit is insufficient to meet all the preheat load.

Example 18.4

The air conditioning system shown in Fig. 18.9 is required to maintain two animal rooms at 21°C for 24 h per day, 7 days a week. The HRU is used as preheater, controlled through thermostat T_1 to maintain a constant temperature of 12°C. Using the design information, determine the annual heating energy saved compared with a heating coil.

Design Data

Air mass flow rate (assumed constant)	2.5 kg/s
Outdoor air design condition	−4°C
Climate data for Heathrow (Table 4.1)	
Boiler firing efficiency (average)	75%

Solution

The load on a preheater is given by Eq. (2.4):

$$Q_h = \dot{m}c_{pas}(t_B - t_A)$$

Since the mass flow rate, specific heat, and efficiency are considered constant, they can be omitted from the calculations until the numerical integration has been completed. The load variations can therefore be expressed in terms of temperature differences.

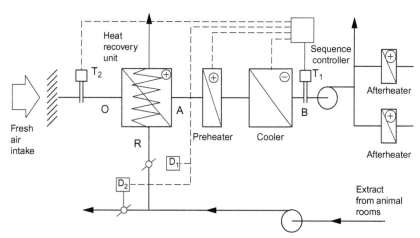

Fig. 18.9 Air conditioning system (system 3), Example 18.4.

The load on the HRU is given by the following:

$$\Delta t_{HRU} = (t_B - t_o) = 12 - t_o \qquad (18.6)$$

The load profile for the operation of the HRU is shown in Fig. 18.10. The numerical integration is completed in Table 18.4.

The accumulated sum of the product is: $\sum (f_1 \Delta t_{HRU}) = 332$.

Using Eq. (18.2), the annual energy saving compared with a heating coil is obtained.

Since $\dot{m}$, c_{pas}, and η are constant:

$$E_c = \frac{\dot{m}_a c_{pas}}{\eta\,100}\,H\sum(f_1 \Delta t_{HRU})$$

$$= \frac{2.5 \times 1.02 \times 8760}{0.75 \times 100} \times 332 = 988,829\,\text{kWh/annum} \; (356\,\text{GJ})$$

This result will be compared with a system using a scheduled temperature at **B**, Example 18.5.

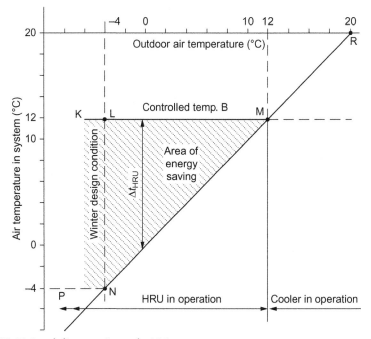

Fig. 18.10 Load diagram, Example 18.4.

Table 18.4 Calculation of energy saving with heat recovery unit in system 3, Example 18.4

Range of outdoor air dry-bulb temperatures (°C)	Midpoint of range t_o(°C)	Frequency f_1(%)	Δt_{HRU}	$f_1 \Delta t_{HRU}$	Notes
<−4	–	0.33	16.0	3.9	Design load
−4.0 to −2.1	−3.0	0.81	15.0	12.2	Use Eq. (18.6)
−2.0 to −0.1	−1.0	2.21	13.0	28.7	$\Delta t_{HRU} = 12 - t_o$
0.0 to 1.9	1.0	4.97	11.0	54.7	
2.0 to 3.9	3.0	7.15	9.0	64.3	
4.0 to 5.9	5.0	9.28	7.0	65.0	
6.0 to 7.9	7.0	11.28	5.0	56.4	
8.0 to 9.9	9.0	12.22	3.0	36.7	
10.0 to 11.9	11.0	11.84	1.0	11.8	
Total	$\sum (f_3 \Delta h_1)$			331.7	

Note: the design load is that of a conventional heating coil.

The estimated annual energy saving can be compared with the capital cost of the HRU and associated ductwork, dampers, and controls. However, account must also be taken of increased fan energy and miscellaneous equipment.

Variations in System Efficiency

In the examples given in this chapter, COP and boiler efficiency have been considered constant. If the variations of these two parameters are known, they can easily be associated with each *bin* and included in the calculations. The technique has been used to investigate the economic application of variable compressor speed for refrigeration system capacity control.

LOAD DIAGRAMS

The variation of heat gains and losses to a building, zone of a building, or an individual room can be plotted against outdoor air temperature in a similar fashion to Fig. 18.10, and the resulting graph is known as a *load diagram* or *load chart*. they are used to assist with system design, planning the economic operation of the plant, and for estimating annual energy consumption.

Referring to Fig. 18.11, the line **AB** represents the transmission heat loss (or gain) $\sum (UA/\Delta t)$; this line may also include heat loss due to air

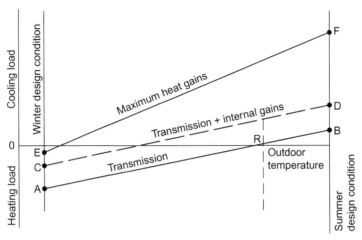

Fig. 18.11 Load diagram for west orientation of a building.

infiltration. Note that **AB** passes through point **R** where the outdoor air temperature is equal to the room temperature t_R and where the transmission loss is zero. It is usual for the design total internal sensible heat gain from occupants, lights, and electric equipment to be considered constant throughout the year. When the total internal heat gain is added to the transmission line, it produces line **CD** parallel to **AB**. To complete the diagram, the maximum solar heat gains calculated at the summer and winter design conditions are added to line **CD** to produce **EF** the line of maximum heat gain (or minimum heat loss). The area **ABFE** represents the heat gain/loss load variations for the space or zone, between the limits of the zone being occupied in summer and the zone being unoccupied at the winter design conditions.

Load diagrams depend on the orientation; Fig. 18.11 is a load diagram suitable for a zone on a west face of a building. For most orientations, **EF** will be a *dogleg* **EFF′** as in Fig. 18.12 where the maximum solar gain does not coincide with the maximum outdoor temperature. For constant air flow rate systems, supply air temperature lines may be drawn in conjunction with the load diagram as shown in Fig. 18.13. The minimum supply temperature, t_S, corresponds to the maximum heat gain at **F**, the temperature difference $\Delta t_C = (t_R - t_S)$, being the design cooling temperature differential used to determine the air flow rate. In a constant air flow rate system, as the room sensible heat gain decreases the supply air temperature increases to give the line **ST**. Similarly, the maximum supply air temperature will correspond to

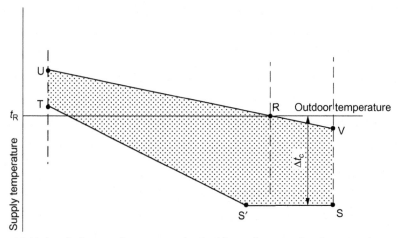

Fig. 18.12 Load diagram for a zone of a building whose peak solar gain does not coincide with the outdoor design temperature.

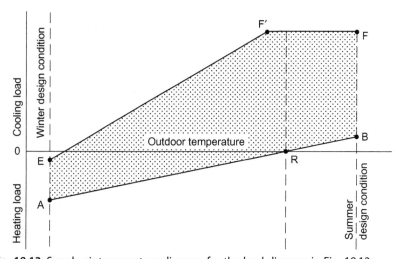

Fig. 18.13 Supply air temperature diagram for the load diagram in Fig. 18.12.

the maximum heat loss at **A** to give the point **U**, whilst the supply temperature line **URV** corresponds to the transmission line at a particular value of outdoor temperature, t_o. The supply air temperature will therefore be at any point located within the area **STUV**, depending on the sensible heat gain or loss to the air conditioned space.

The energy consumption of most systems will depend on the frequency of occurrence of the loads within the load diagram. The BIN method would allow this to be done relatively easily by incorporating within the diagram the predicted frequency of occurrence of the heat gains within each *bin*. The corresponding frequency of occurrence of the air supply temperatures for each of these loads can then be obtained.

The method given in the ASHRAE Guide is to produce a *mean load line*. Here, the heat gains are multiplied by load diversity factors based on occupancy levels, use of heat generating equipment, and solar heat gains; typical figures for the United Kingdom are 40% for summer months and 17% for winter design months, but since these figures underestimate the effect of indirect radiation, figures of $k_{s1} = 0.5$ and $k_{s2} = 0.2$ might be used.

This is illustrated in Fig. 18.14 where average load lines have been added to the load diagram of Fig. 18.11. Line **GH** is obtained by adding the average internal gains to the transmission line:

$$q_{AG} = q_{BH} = k_i q_{BD}$$

where k_i is the internal heat gain load diversity factor.

The average heat gain at the summer design condition is obtained:

$$q_K = q_H + k_{s1} q_{DF}$$

The average heat loss at the winter design condition is obtained:

$$q_J = q_G + k_{s2} q_{CE}$$

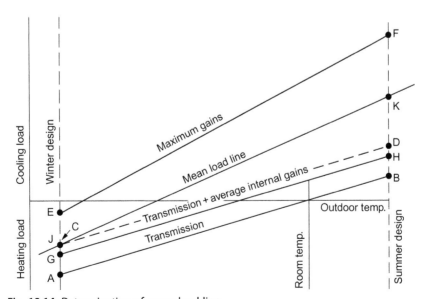

Fig. 18.14 Determination of mean load line.

The mean load line is arrived at by joining points **J** and **K**. Note that **JK** extends beyond the design conditions when the outdoor temperatures are more severe than the design conditions.

SCHEDULING

To improve the operating efficiency of some systems, temperatures within the plant can be scheduled to the outdoor air dry-bulb temperature. The schedules are determined from an analysis of the load and temperature diagrams discussed above. The analysis can also include the calculation of the energy consumption, and this is illustrated by the following example of the air-to-air HRU in the 100% system in Fig. 18.9.

Example 18.5
Referring to Fig. 18.9, the output of the HRU is regulated through thermostat T_1, reset by T_2 in the fresh air duct, to maintain a scheduled temperature at **B**. Using the design data, determine the annual heating energy saved by using a HRU in place of a traditional (hot water or steam) preheater.

Design Data

Mass air flow rate (assumed constant)	2.5 kg/s
Outdoor air design condition	$-4°C$
Climate data for Heathrow, Table 4.1	
Boiler firing efficiency, η	75%

The temperature rise across the HRU related to the outdoor dry-bulb temperature, to, by the equation

$$\Delta t_{HRU} = 14 - 0.7\,t_O \qquad (18.7)$$

The precalculated temperature schedule line **US'** (based on the relevant load diagram) is given by the equation

$$t_s = 19 - 0.4\,t_O \qquad (18.8)$$

Solution
The temperature analysis is prepared on Fig. 18.15. Line **US'S** is the variation of minimum supply temperatures required to meet maximum heat gains of the load diagram and is the *schedule* line. Line **VPR** is the dry-bulb temperature of the outdoor air entering the system. Line **LNR** is the temperature, t_a, based on Eq. (18.9). To maintain the schedule line **US'S**, the output of the HRU has to be reduced from point **N** to point **P**. The heating requirements of the HRU are then represented by

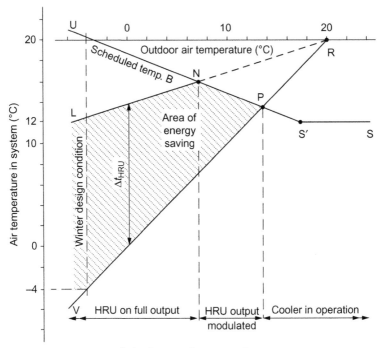

Fig. 18.15 Temperature analysis diagram for system 3.

the shaded area **LNPV**; this will give the energy saving compared with a conventional preheater.

Since the mass flow rate, $\dot{m}_a$; specific heat, c_p; and boiler efficiency η are considered constant, they are omitted temporarily from the calculations until the numerical integration has been completed. The load variations can then be expressed in terms of temperature differences.

At the outdoor air conditions between -4 and $10°C$, the temperature rise would be that for a conventional heater and Eq. (18.6) applies. Between 7 and $14°C$, the temperature rise across the HRU will be the difference between line **UNP** and the outdoor temperature.

Equation of line **UNP**:

$$t_s = 19 - 0.4\,t_O$$

$$\Delta t_{HRU} = t_s - t_O$$

$$\therefore \Delta t_{HRU} = (19 - 0.4\,t_O) - t_O$$

$$\Delta t_{HRU} = 19 - 1.4\,t_O \tag{18.9}$$

The numerical integration is completed in Table 18.5.

Table 18.5 Calculation of average hourly energy saving of heat recovery unit in system 3 (Example 18.5)

Range of outdoor air dry-bulb temperatures (°C)	Midpoint of range t_1 (°C)	Frequency f_1 (%)	Δt_{HRU}	f_1 Δt_{HRU}	Notes
<-4.1	—	0.33	18.0	5.9	Design load
-4.0 to -2.1	-3.0	0.81	16.4	13.2	Use Eq. (18.7)
-2.0 to -0.1	-1.0	2.21	14.7	32.5	$\Delta t_{HRU} = 14 - 0.7\,t_o$
0.0–1.9	1.0	5.97	13.3	79.4	
2.0–3.9	3.0	7.15	11.9	85.1	
4.0–5.9	5.0	9.28	10.5	97.4	
6.0–6.9	6.5	5.46	9.5	51.9	
7.0–8.9	8.0	12.03	7.8	93.8	Use Eq. (18.9)
9.0–10.9	10.0	12.09	5.0	60.5	$\Delta t_{HRU} = 19 - 1.4\,t_o$
10.0–12.9	12.0	11.39	2.2	25.1	
13.0–13.5	13.2	3.2	0.5	1.6	
Total	$\sum (f_3\, \Delta t_1)$			546.4	

Note: design load is that of a conventional heating coil.

The accumulated sum of the product is $f_1\Delta t_{HRU} = 546$.
Using Eq. (18.2), the annual energy consumption is obtained:

$$E_c = \frac{H}{100}\sum\left(\frac{fQ}{\eta}\right)$$

Since $\dot{m}_a$, c_{pas}, and η_b are constant:

$$E_c = \frac{\dot{m}_a\,c_{pas}\,H}{\eta_b \times 100}\sum(f_1\Delta t_{HRU})$$

$$= \frac{2.5 \times 1.02 \times 8760}{0.75 \times 100} \times 546 = 162,621\,\text{kWh/annum}\,(585\,\text{GJ})$$

Comparing this result with the unscheduled system in Example 18.4, the scheduled system shows a 64% increase in the preheater energy saved by using an HRU.

FAN AND PUMP ENERGY CONSUMPTION

The annual cost of electric energy for a fan or a pump driven by an electric motor at constant speed and supplying a constant volume flow rate is calculated from the equation:

$$E_F = \frac{\dot{V}\,P_{tF}H}{\eta_F\,1000} \tag{18.10}$$

where suffix F denotes either fan or pump.

The SI units used to specify fan and pump duties will not always be identical. Those applying to Eq. (18.9) would be the following:

Volume flow rate, $\dot{V}$ m/s

Pressure, P_{tF}, Pa

Where air flow rates vary, as in a VAV system, the BIN method can be used; in which case, Eq. (18.10) must also include the efficiency of the fan speed control.

SYMBOLS

c_{pas} specific heat of humid air
E_c annual energy consumption, cooler
E_F annual energy consumption of fan or pump
E_h annual energy consumption, heater
f_1 frequency of occurrence of outdoor air dry-bulb temperature
f_3 frequency of occurrence of outdoor air enthalpy
H annual hours of plant operation

H_{eq} equivalent annual hours of plant operation
h specific enthalpy
k_i internal heat gain load diversity factor
k_{s1} solar heat gain load factor, summer
k_{s2} solar heat gain load factor, winter
$\dot{m}_a$ air mass flow rate
Q heat flow rate
q room heat gain/loss
P_{tF} fan total pressure
t air temperature
$\dot{V}$ volume flow rate
x fraction of outdoor air
Δh enthalpy difference
Δt temperature difference
η_b boiler firing efficiency
η_F fan efficiency

SUBSCRIPTS

B   cooler off-coil condition
d   design load
M   mixed air condition
O   outdoor air condition
p   preheater
R   room air condition
S   supply air condition

ABBREVIATIONS

COP   coefficient of performance
HRU   heat recovery unit
VAV   variable air volume

REFERENCES

[1] Degree Days: Theory and Application, CIBSE TM41, 2006.
[2] ASHRAE Handbook—Fundamentals, 2013.

CHAPTER 19

Commissioning, Operation and Maintenance

Beginning in itself has no value; it is an end which makes beginning meaningful; we must end what we begun.

Amit Kalantri

COMMISSIONING

As air conditioning systems and other building engineering services have increased in size and complexity, it is important not only that they work but that they operate efficiently. To bring the installed systems into full working order they have to be *commissioned*. Commissioning in the context of building engineering services is a term used in the United Kingdom for the process of bringing the systems into use and ensuring that the systems comply with the specification before they are handed over to the client. (The *ASHRAE Handbook* uses the term *testing, adjusting, and balancing* [1]).

The CIBSE code for air-distribution systems [2] gives the following definition of commissioning (other codes have similar definitions):

The advancement of an installation from the stage of static completion to full working order to the specified requirements. It includes the setting to work and regulation of the system flow rates.

Static completion is the state of the system installed in accordance with the specification, ready for *setting to work*, which includes cleaning and pressure testing the distribution system, where required.

PROCEDURES

Commissioning should be viewed as a continuous process during the project cycle. The interdependence of the various stages in the project cycle and the requirements of the building and other services should be recognized and the work planned accordingly. In many ways, commissioning can be considered

Air Conditioning System Design
http://dx.doi.org/10.1016/B978-0-08-101123-2.00019-4

as a management process in that it requires good organization, documentation, and definition of responsibilities for each part of the work.

Notes for each stage of the project cycle are given below, in order that some of the technical aspects can be seen in perspective, the emphasis placed on the procedures for the air-handling side of air conditioning. The approach for other services associated with air conditioning, e.g., boiler plant, automatic controls, refrigeration systems, and water distribution, is given in the relevant commissioning codes.

The system(s) has to be set to work, balanced, and tested before the installation is accepted by the client as complete and satisfactory, but it should be noted that some testing could be carried out during the guarantee period, after the plant has been handed over to the client, as indicated later in the chapter.

Design Requirements

A system that operates to the design requirements depends initially on sound design; the commissioning of an installation will not make a poor design operate efficiently, but will only highlight any shortcoming and failures. More specifically, the system must be designed for the commissioning procedures that follow the completed installation. This means that the facilities for measuring and regulating the variables have to be incorporated in the design and specification, to ensure that the work entailed is carried out efficiently and effectively.

The ideal will be a self-balancing system, so that when the system is started up, each outlet or unit will deliver its design output. This approach can be more expensive than a system that requires manual adjustment, but the additional cost could well be offset by the reduced time required for setting the systems to work, the benefit of long-term stability and fewer maintenance problems. Self-balancing systems are not always a practical solution, but there is scope in the arrangement of duct and piping layouts to ensure easier regulation of the flow rates to achieve the system balance required by the design.

Decisions will often have to be made at the conceptual stage of a project so that the detailing of the design can proceed. An understanding of the balancing procedures described in Chapter 15 is essential to the design requirements.

The following is a list of the more important provisions required; some of these facilities will be those required for maintenance:

- Measurement stations for system variables, e.g., flow rates, pressures, and temperatures;
- Instrumentation for balancing, performance tests, and efficient operation;
- Regulating devices, e.g., dampers and valves;
- Access and inspection openings;
- Filling, drainage, air venting, and flushing of water circuits;
- Electric provisions for convenience, safety, and isolation of subsystems.

Construction and Installation

No matter how complete the design and specification, there will always be on-site problems. Adequate supervision is essential to ensure correct installation and good liaison between the designer's office, the contractor's office, and the site. The main contractor and specialist subcontractors should act together so that the work can be adequately planned and executed with a member of the design team assisting to ensure the correct interpretation of the design.

Supervisors representing the design engineer and contractor should be appointed, and they must be supplied with the design information. To ensure that all plants are adequately constructed and installed, checklists are used to ensure systematic inspections.

Any faults that are found can be summarized (a *snagging list*) for action by the contractor, and a later check made to verify that they have been rectified. It is essential that these inspections are carried out early enough, since, for example, it would be impossible to inspect adequately ducts covered with builder's work or hidden by other services. It may also be necessary to witness tests on certain plant items at the factory, before delivery to site. There should be adequate physical protection of plant that has to be installed well before being made operational. In these ways, a smooth transfer to the stage of *setting the plant to work* can be achieved with a minimum of backtracking.

Setting the Plant to Work

Prestart-up checks will be required to ensure that the plant is complete, clean, and safe to operate. These checks are listed in the various codes and are usually in addition to those inspections carried out during installation. Once the plant has been found to be physically correct and to have the necessary supply services available, it can then be systematically started up and balanced. The procedure would be tailored to suit individual projects, but a typical approach for an air conditioning system would be as follows:

- Water circuits made operational, including initial setting to work of the heating and cooling plants;
- Inspection of plant under running conditions;
- Air circuit balancing;
- Water circuit balancing;
- Final checks on boiler and refrigeration plants;
- Control calibration.

It will usually not be possible to complete each stage before moving on to the next. There will naturally be a good deal of overlapping between the various phases, and it will take experience and good planning to ensure a satisfactory sequence of activities.

Water Circuits Operational

The commissioning of boiler plant and heating circuits is a priority if warm air is required during the cold months of the year. To have chilled water available whilst the air circuits are being balanced will not normally be necessary, though the commissioning of the refrigeration plant can also commence at this stage.

Flushing the piped water systems through to ensure cleanliness is of utmost importance to the eventual satisfactory performance of the system.

Inspection of Plant Under Running Conditions

Checklists should be available similar to those required for the static inspections. A typical list covering several components in an air-distribution system is shown in Fig. 19.1 [3]. It may be necessary to bring in manufacturers for specialist plant items, but a *system* approach is important in order that the performance as a whole can be considered.

Air and Water Circuit Balancing

The balancing procedures for air and water systems have been described in Chapter 15. The regulation of water distribution systems by thermal methods is excluded from the CIBSE Code A as it is considered to be unreliable and wasteful of commissioning time (as is also the case for systems with a close approach of air and water temperatures, such as chilled water systems). However, if the water circuits are associated with heaters in an air-conditioning system, then the measurements can sometimes be simplified by using the rise in air temperature across the heaters to regulate the water flow rates. When this course is adopted, it implies that design air flow rates

5. AIR DISTRIBUTION SYSTEM – SETTING-TO-WORK CHECKLIST		**BSRIA**
Client:		
Project:		
System:		www.bsria.co.uk

Check that:	✓/✗	Comments / Follow-up references
Prior to fan start		
1. All branch regulating dampers open		1.
2. Fan main damper 50% open		2.
3. All fire dampers open		3.
4. Automatic control dampers set for full fresh air or full recirculation		4.
5. Grille louvres set square to face		5.
6. Ceiling diffuser cones set for full downward		6.
7. Supply and extract fans running at same time		7.
8. All windows and doors in the building closed		8.
9. Installer ready to start system		9.
Initial start		
10. Direction and rotation speed of motor shaft is correct		10.
11. Motor, fan, and drive are free from vibration and undue noise		11.
12. Motor starting current is correct for sequence timing adjustments		12.
13. Motor running current is balanced between phases		13.
14. There is no sparking at the commutator or slip rings		14.
15. Motor and bearings are not overheating and water coolant is adequate		15.
16. There is no seepage of lubricant from the housing		16.
17. Reduced speed and motor running currents are correct on multi-speed motors		17.
Initial run		
18. Fuses, switchgear, and motor are not overstressed		18.
19. Motor current reaches design value or full load current, whichever is the lower		19.
20. Fan pressure developed does not exceed system design pressure		20.
Running-in period		
21. Bearings and motor temperature remain steady		21.
22. Gland nuts are adjusted to give correct drip rates		22.

GENERAL COMMENTS

Date:	Engineer:	Approved by:	Sheet:

Fig. 19.1 Plant performance summary checklist. *(Reproduced with permission of BSRIA, BSRIA BG 49/2015.)*

should first be obtained by balancing the air flow circuits. To obtain full output from the heater, control valves should be fully open to load; this can be achieved by a manual adjustment of the thermostat setting.

Automatic Controls

Automatic controls are covered by the CIBSE Commissioning Code C. This code is of a general nature, and reference to the specialist manuals prepared by the controls equipment manufacturer is recommended.

Though some prestart-up checks can be made prior to setting to work the remainder of the system, final calibration of the control systems can be completed only after the regulation of the air, water, boiler, and refrigeration systems.

It is probable that final tuning to obtain maximum efficiency is possible only after the building has been occupied, and the building and systems have had any teething troubles rectified. The client and/or the building occupier must be aware of this when contracts are placed and responsibilities defined.

ORGANIZATION OF COMMISSIONING

Commissioning as defined by the CIBSE includes the setting to work and regulation of the flows and the calibration of the controls. *Setting to work* is the process of bringing a static system into motion, and regulation is the process of adjusting the rates of fluid flow in a distribution system within specified tolerances required by the design. Commissioning does not replace the inspections required during the installation of the plant nor is it an optional extra. Adequate documentation and design information must be available to the personnel responsible for this section of the work. Records of the measurements and adjustments to the system must be kept at the time they are made, as commissioning is often an intermittent operation; these records will show the state of progress at any stage. Records are also necessary to provide a base for comparing the state of the system during maintenance and operation.

For successful commissioning, it is important that the responsibilities of each engineer involved are clearly defined and incorporated in contracts and terms of appointment. The responsibility for deciding by whom each section of the work should be done generally rests with the design engineer. Several alternatives are available, the most appropriate of which should be selected for each project. These include the following:

1) *Design and site engineers.* The designer will decide how variables are to be measured so that the necessary facilities are built into the installation. The site engineer will be responsible for inspections, setting the plant to work and regulating the flow circuits. The designer or the client's representative will then verify that the installation is correct and satisfies the design intent, by either examining records or making spot checks. The advantage of this arrangement is that the designer and site engineer, being intimately connected with the installation, will know the location of all equipment and controls. The designer will know what is expected of the plant and will be able to recognize and quickly correct any faults that occur.

2) *Manufacturers.* It is common practise for suppliers of specialized equipment (such as boilers) to commission their own plant. This idea could be extended to cover practically all equipment, but to be effective, cooperation, goodwill, and coordination between all parties are essential.

3) *Commissioning teams.* Some firms have their own commissioning teams who carry out all the commissioning and performance testing on plant items, even if it was designed and installed by an outside contractor. Independent, commercial commissioning teams are also used. The advantages are that the commissioning can be done quickly and efficiently, with all the necessary instruments to hand, and that design-in-use surveys can be carried out. The design and site engineers will be freed earlier to concentrate on their own specialities. At the design stage, commissioning engineers will be able to give up-to-date advice on all provisions to be made for the measurement and regulation of variables.

PERFORMANCE TESTING

The commissioning process should ensure that the plant is set to work in a methodical manner and balanced to the design requirements. That is, water, heating, refrigeration, and electric circuits are made operational, and equipment is started and inspected under running conditions. Air volume flow rates are regulated to ensure specified output for heaters, coolers, and humidifiers, controls calibrated to maintain the design conditions in the building. This commissioning process will not necessarily show that the plant is capable of operating satisfactorily throughout the year; to do this, a certain number of performance tests should be applied to the plant.

The major problem encountered in ensuring the plant will satisfy the building requirements at the summer and winter design conditions is that when the plant is set to work, it will rarely be operating at these conditions. In fact, outdoor design conditions may only occur, say, once in 10 years. Most tests must therefore be carried out at off-peak conditions and a prediction made from the test results of the ability of the plant to maintain the indoor conditions when the outdoor design conditions are reached. Other problems are that after the commissioning process has been completed, the building has not had time to *settle down*, systems may not be subject to normal occupancy loads, and the client will want to occupy the building with a minimum of delay.

It is important to distinguish between two types of tests associated with plant installed in a building. One of these is to ascertain whether the plant delivers its rated output; the second is to ascertain whether the installed plant will satisfy the building's requirements. The first of these is a check on the manufacture's data, with the plant item working with other components; the second is to check on the designer's own calculations.

At present, there is no agreement within the building services industry as to what performance tests, if any, should be adopted. The CIBSE Commissioning Codes have defined *testing* as follows:

> The measurement and recording of system parameters to assess specification compliance.

However, the type of test suggested here is not covered by the codes and is a separate consideration. Though some tests are carried out in the commissioning process, no attempt has yet been made to produce documentation for system testing as defined above. BS 5720:1979 [4] recommended that performance testing should be carried out, listing some that may be appropriate:

- Heating capacity to satisfy total energy demands for humidifying and heating loads;
- Zone heater/cooler capacity to meet space heating/cooling loads;
- Plant cooling capacity to meet total energy demand for dehumidification and cooling loads.

Before testing commences, checks should be made that plant items function correctly; a faulty control valve or a ghost circulation,[1] for instance, could invalidate a test.

---

[1] See Appendix B.

Discussion

If plants are to perform as required by the designer and client and if future designs are to be improved, it is desirable that tests are carried out under actual operating conditions, usually with the building occupied. This will undoubtedly raise a number of contractual issues that need to be resolved within the guarantee or fault liability period.[2]

Other methods of test have been suggested and used from time to time, e.g., heaters can be installed in the spaces to load up the cooling system; a preheater can produce a load for the cooling coil. Control settings can be adjusted to obtain design temperature differences, when the external conditions are not at the design values. Though there might be some merit in these methods, it still has to be demonstrated that the plant will operate satisfactorily under actual running conditions, and artificial loading of the plant/building may not be able to do this.

The ability of the plant to meet design conditions during the life of the plant will depend on adequate maintenance. If regular tests are carried out as part of planned maintenance, failures and deteriorations can be detected by comparing the results with the original tests.

With present-day instrumentation, data logging through BMS systems, and computer techniques using mathematical models of the system performance, testing should gain greater acceptance.

MAINTENANCE

Design for Maintenance

As with commissioning, decisions will often have to be made at the conceptual stage of the project so that the installed system can be operated and maintained efficiently and safely. The following is a list of some of the more important aspects to be considered:

1) *Plant rooms.* Of prime importance is the provision of adequate building space for the installation of equipment and for its regular servicing, as well as for the possible removal, replacement, and repair of major plant items. Plant rooms should be adequately heated and ventilated.

---

[2] The British Government has recognized the need for contractors to be engaged in Whitehall projects after handover with more time for the commissioning process. In addition, project teams will be expected to appoint a 'GSL champion' to ensure end-user engagement in the design and construction process. (*CIBSE Journal*, July 2013). This idea could be extended to other clients.

2) *Lifting and strong points.* These should be provided, where necessary, to allow dismantling and assembling and possible removal and replacement of major plant items.

3) *Access.* Reasonable access is required to all major plant items and most distribution ducts. This provision is required both in the building work chambers, ducts, and false ceilings, as well as in the plant and ductwork system themselves. Access *drawings*, akin to builders work drawings, are useful for highlighting this important aspect of maintenance. Such drawings also identify access *routes* through the building; these routes are for delivery of large plant items at the installation stage and for the subsequent removal and replacement of equipment.

4) *Instrumentation.* To test the systems for efficient operation, certain instruments will be needed. Such instruments must be certified and tested for accuracy.

5) *Standby plant.* Care is needed in providing adequate standby plant to ensure the system as a whole continues to operate, when required, in the event of a component failure. An overgenerous provision means incurring unnecessary capital cost, with plant standing idle and depreciating in value. Partial standby can often be achieved by selecting plant items having more than one stage to meet the total capacity.

6) *Spares.* Spare parts for essential repairs, together with replacement items such as clean filter cells, are required.

7) *Safety features and isolation of subsystems.* Systems should be easy to operate, understandable, and safe. Adequate provision is required to minimize fire risk and to isolate electrical circuits. Various interlocks and alarms will be required with boilers, pressure vessels, refrigeration plant, and fans. The requirements to protect personnel and equipment will follow normal design practise, e.g., vents, isolating valves, pressure switches, and safety guards.

8) *Materials.* The materials of each item of plant, duct, pipe, insulation, and so on must be compatible with and suitable for the conditions under which the system will be operating.

Hand-Over and Documentation

Designers and contractors can help to ensure adequate maintenance by stressing its importance to the client. The client should be advised about the personnel required and provided with the requisite information for maintenance.

For plant to continue working satisfactorily throughout its life, it is necessary for the maintenance and operating staff to be familiar with the principles and methods of running the plant. In addition to a full set of record drawings, two manuals should be available, one for the operating staff (e.g., office manager and caretaker) and another for the maintenance staff. The former should describe such things as the design principles, the method of operation, details of alarms, and safety precautions. The latter, a larger service manual, should contain the information listed by BS 5720:1979 [5] and reproduced below.

Documentation Required for System Maintenance Engineer

- The designer's description of the installation, including simplified line flow and balance diagrams for the complete installation;
- As fitted installation drawings and the designer's operational instructions;
- Operation and maintenance instructions for equipment, manufacturer's spare parts lists, and spares ordering instructions;
- Schedules of electric equipment;
- Schedules of mechanical equipment;
- Test results and test certificates as called for under the contract, including any insurance or statutory inspection authority certificate;
- Copies of guarantee certificates for plant and equipment;
- List of keys, tools, and spare parts that are handed over.

British Standard 5720:1979 [6] provides further advice regarding the organization and content of maintenance manuals. These should be available in draught form for checking at the commissioning stage, in addition to *as fitted* drawings. This will assist those concerned with setting the plant to work efficiently, and at the same time, the manual can be revised to suit any operational changes that may have been necessary, before they are issued to the client/building owner.

It is important that before the plant is handed over, the staff responsible for operating the plant is given verbal instruction and demonstrations on the principles and operation of the systems. Any such verbal instructions should be in addition to the documentation.

Maintenance Organization

Once the client has accepted an installation from the contractor, the maintenance of a plant of any complexity should be organized to ensure continued efficient operation of the plant, aiming to protect the capital investment at a minimum economic cost.

The basis of any planned maintenance scheme is a system whereby the checks and services to be carried out on any piece of plant come to light at the appropriate time. When staff completes a piece of maintenance, this should be recorded and anything in need of attention (or likely to be in need of attention in the near future) such as a bearing running hot that would eventually fail. This enables the repair to be carried out at a convenient time, rather than in a period when everything seems to fail at once.

Contract maintenance by specialist firms is often used as an alternative to directly employed labour, either for a part or the whole of the service.

Frequency of Servicing

Routine maintenance includes inspections, cleaning, water treatment, adjustment, and overhaul. The frequency at which these should be made is normally given in the manufacturers' manuals, but these are average values that are best modified by the actual site conditions and in the light of operating experience.

The frequency at which plant is serviced depends on the following:
- Plant and system efficiency and hence efficient energy consumption;
- Effect on reliability of service;
- Routine maintenance costs;
- Fault repair costs;
- Safety inspection;
- Hours of system operation.

As part of the routine maintenance inspections, standby and emergency plant must be checked but not necessarily brought online for long operating periods.

Following shutdowns for repairs and plant modifications, it may be necessary to recommission the system or part thereof, in which case the appropriate procedures should be followed.

It is particularly important to include insurance inspections of pressure vessels and the testing of fire alarms in routine maintenance.

Fault-Finding

Fault-finding procedures may be included in the service manual. Though these procedures for individual plant items will often be available from manufacturers, they should also relate to the system in which the plant item is placed.

Maintenance Support

In support of maintenance, it is recommended that consideration be given to the following:

- Engineer's office;
- Workshop with appropriate tools;
- Equipment spare parts;
- Maintenance materials;
- Instruments;
- Site tools.

Operating Efficiency

An important objective of maintenance is to operate the systems as efficiently as possible after commissioning has been completed. In this respect, there is an overlap between commissioning and maintenance. It is not always possible to obtain full system efficiency prior to the occupation of the building, and it is probable that maximum efficiency can be achieved only with the building in normal use. The building use may change after first occupancy, and this may also affect the optimum plant operation.

If the system is not operating as efficiently as it should or if economies need to be made, an energy audit of the plant may be undertaken; this would indicate where to place most emphasis to achieve energy savings. For normal performance monitoring, it is necessary that maintenance staff have as good an understanding as possible of the overall and detailed system design.

Risk Assessment

All systems require a risk assessment. It is important to consider the system as a whole and not, for example, a cooling tower in isolation (e.g., re *Legionella*).

REFERENCES

[1] ASHRAE Handbook, HVAC Applications 2014.
[2] CIBSE Commissioning Code A: Air Distribution Systems, 2004.
[3] Commissioning Air Systems, BSRIA BG 49/2015.
[4] BS 5720: 1979, Mechanical ventilation and air conditioning in buildings, Code of Practice, British Standards Institution (withdrawn, though still available).
[5] BS 5720: 1979, op.cit. p 60.
[6] Ibid., p 80.

APPENDIX A

Derivation of Equations for Inherent and Installed Characteristics of a Flow Control Device

INHERENT CHARACTERISTIC

The inherent characteristic γ' is defined by the relationship between the stroke of the flow control device (FCD) (or angular position of the damper blades) and the flow rate relative to the maximum flow in the fully open position whilst maintaining constant pressure drop across the device. It is given by the equation:

$$\gamma = \frac{\dot{v}_\theta}{\dot{v}_\varnothing}$$

If the pressure drop across the control device is defined in terms of pressure loss coefficients, then:

$$\Delta p = K \, 0.5 \, \rho \bar{v}^2$$

From the definition of the inherent characteristic:

$$\Delta p_\varnothing = \Delta p_\theta$$

$$\therefore K_\varnothing \, 0.5 \, \rho \bar{v}_\varnothing^2 = K_\theta \, 0.5 \, \rho \bar{v}_\theta^2 \qquad (A1)$$

If A is a representative area of the device, then:

$$\bar{v} = \frac{\dot{v}}{A}$$

The inherent characteristic is then given by:

$$\gamma = \frac{\dot{v}_\theta}{\dot{v}_\varnothing} = \frac{\bar{v}_\theta}{\bar{v}_\varnothing}$$

Therefore, from Eq. (A1),

$$\gamma = \sqrt{\frac{K_\varnothing}{K_\theta}} \qquad (17.2)$$

In a similar way, the inherent characteristic can be expressed in terms of resistances, i.e.,

$$\gamma = \sqrt{\frac{r_\varnothing}{r_\theta}} \qquad (A2)$$

INSTALLED CHARACTERISTIC

The installed characteristic γ' is defined by the relationship between the stroke of the FCD and the flow rate through the system relative to the flow rate with the FCD in the fully open position, i.e.,

$$\gamma' = \frac{\dot{v}_\theta}{\dot{v}_\varnothing}$$

Consider the general case where the total pressure available in the system varies with the flow rate. A typical example of this is where the control device is used to *throttle* the total flow rate of a fan or pump. If the prime mover characteristic is defined by Eq. (15.7):

$$p_F = A' - B'\dot{V}^2 \qquad (A3)$$

At maximum flow rate and with the FCD fully open ($\theta = \varnothing$), the total system pressure loss Δp_{st} can be expressed as:

$$\Delta p_{st} = \Delta p_s + \Delta p_\varnothing \qquad (A4)$$

where Δp_s is pressure drop in the system excluding the pressure drop across the FCD and $\Delta p_\varnothing$ is the pressure drop across the fully open FCD.

Using resistances in series, given by Eq. (13.17):

$$\Delta p_{st} = r_{st}\dot{V}^2_\varnothing + r_\varnothing \dot{V}^2_\varnothing \qquad (A5)$$

At the operating point, $p_F = \Delta p_{st}$. Therefore, equating (A3) and (A5) and with the FCD fully open at stroke at the maximum flow rate:

$$A' - B'\dot{V}^2_\varnothing = r_{st}\dot{V}^2_\varnothing + r_\varnothing \dot{V}^2_\varnothing \qquad (A6)$$

At the reduced flow rate with the FCD at position θ:

$$A' - B'\dot{V}^2_\theta = r_{st}\dot{V}^2_\theta + r_\theta \dot{V}^2_\theta \qquad (A7)$$

Therefore, from Eqs (A6), (A7), the installed characteristic is obtained as:

$$\gamma' = \frac{\dot{V}_\theta}{\dot{V}_\varnothing} = \sqrt{\frac{r_{st} + r_\varnothing + B'}{r_{st} + r_\theta + B'}}$$

With the inherent characteristic expressed in terms of resistances, from Eq. (17.14):

$$r_\theta = \frac{r_\varnothing}{\gamma^2}$$

The FCD authority N is defined by Eq. (A2). Therefore:

$$N = \frac{\Delta p_\varnothing}{\Delta p_s + \Delta p_\varnothing} = \frac{r_\varnothing}{r_s + r_\varnothing} \tag{A8}$$

$$\therefore \gamma' = \sqrt{\frac{r_\varnothing \left(\frac{1}{N} - 1\right) + r_\varnothing + B'}{r_\varnothing \left(\frac{1}{N} - \frac{1}{\gamma^2}\right) + B'}}$$

$$\gamma' = \sqrt{\frac{r_\varphi / N + B'}{r_\varphi \left(\frac{1}{N} + \frac{1}{\gamma^2}\right) + B'}} \tag{17.3}$$

For a system with a constant pressure drop, $B' = 0$. Therefore:

$$\gamma' = \frac{\gamma}{\sqrt{[N + \gamma^2(1 - N)]}} \tag{17.4}$$

SYMBOLS

A', B'	constants of fan characteristic
A	representative area of FCD
a	intercept of Eq. (17.5)
b	slope of Eq. (17.5)
D	plug diameter
G_d	damper sizing constant (defined by Eq. (17.9))
K	pressure loss coefficient (defined by Eq. (13.12))
N	authority
P_F	fan total pressure
r	resistance
$\dot{V}$	volume flow rate
$\bar{v}$	mean velocity
γ	inherent characteristic

γ' installed characteristic
Δp pressure drop
θ damper blade angle or position of valve stroke
$\varnothing$ damper start angle or valve fully open

SUBSCRIPTS

d   damper
s   system pressure loss excluding FCD
st   system total pressure loss
θ position of damper blade angle or valve stroke
$\varnothing$ initial position of start angle of damper blades or valve stroke (damper or valve fully open)

APPENDIX B

Ghost Circulations in Pipework Systems

Incorrect piping connections in a pumped circuit can cause unnecessary heating and cooling loads, resulting in reduced efficiency. These are sometimes known as 'ghost circulations'.

As an example, consider a preheater followed by a cooler that are controlled in sequence by a thermostat, as shown in Fig. B.1. When the cooler is

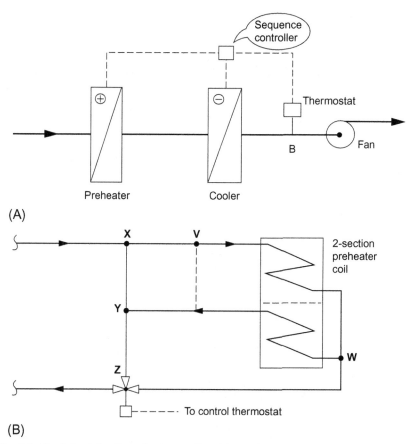

Fig. B.1 Ghost circulation through a two-section air heater battery.
(A) Plant arrangement (B) Piping arrangement of preheater.

in operation, the heater should not provide heat; if it does operate due to a fault in some part of the system, then additional cooling must be supplied to maintain condition **B**. Hence, both heating and cooling circuits are working inefficiently, even though the design conditions are being maintained, i.e., by thermostat at **B**.

The ghost circulation is caused in the two-section heater in the following way. The three-way control valve is closed to heater and open to bypass. Flow is therefore along **XW** causing a pressure difference between **X** and **Y** and hence circulation along the path **XVWY**. The correct piping connection should be made at point **V** instead of **Y**.

These unwanted ghost circulations can be recognized through suitable checks during inspection of plant under running conditions.

APPENDIX C

SI Units

All quantities in this book are given in the SI system of units, which is based on six units of measurement, i.e.,

1) length—meter (m)
2) mass—kilogram (kg)
3) time—second (s)
4) electric current—ampere (A)
5) temperature—degree Kelvin (K)
6) luminous intensity—candela (cd)

From these are derived the remainder of the units necessary for measurement, e.g., area from length (m^2) and velocity from length and time (m/s). Special units are given to some of these derived units, as follows:

Quantity	Unit	Symbol	Basic units involved
Frequency	hertz	Hz	$1\ Hz = 1c/s$ (1 cycle per s)
Force	newton	N	$N = 1\ kg\ m/s^2$
Work, quantity of heat, and energy	joule	J	$1\ J = 1\ Nm$
Power	watt	W	$1\ W = 1\ J/s$
Pressure	pascal	Pa	$1\ Pa = 1\ N/m^2$

Multiples of SI units are increased or decreased by the use of named prefixes, each of which has an agreed symbol. Those most relevant to this book are given in the table below. (Note that kilogram, which is a basic unit, departs from the general rule.)

Multiplying factor	Prefix name	Prefix symbol
10^9	giga	g
10^6	mega	m
10^3	kilo	k
10^{-3}	milli	m
10^{-6}	micro	μ

CONVERSION FACTORS

The conversion factors given in the following tables are for those physical quantities most commonly used in air conditioning, mechanical ventilation, and refrigeration. To convert a quantity in *British units* to *SI units*, multiply by the conversion factor; to convert a quantity in *SI units* to *British units*, divide by the conversion factor.

Physical quantity	SI unit		British unit	Conversion factor
	Description	Symbol		
Space				
length	meter	m	foot	0.3048
			inch	25.4
area	square meter	m^2	square foot	0.09
volume	cubic meter	m^3	gallon (the United States)	0.0378
			gallon (the United Kingdom)	0.0455
Mass				
mass	kilogram	kg	pound	0.454
			ton	1016
Moisture content				
moisture content	kilogram per kilogram	kg/kg	grains per pound	1.43×10^{-4}
Density				
density	kilogram per cubic meter	kg/m^3	pound per cubic foot	16.02
Velocity				
velocity	meter per second	m/s	foot per minute	5.08×10^{-3}
Flow rate				
volume flow rate[a]	cubic meter per second	m^3/s	cubic foot per minute (cfm)	0.472
mass flow rate	kilogram per second	kg/s	pound per hour	1.26×10^{-4}

—cont'd

Physical quantity	SI unit Description	Symbol	British unit	Conversion factor
Pressure				
pressure[b]	pascal	Pa	inch water gauge	249.1
Temperature				
scale, zero 0°C	degree Celsius	°C	degree Fahrenheit (°F)	$\dfrac{5(°F-32)}{9}$
interval	degree Kelvin	K	degree (F)	0.56
Viscosity				
viscosity (kinematic)	square meter per second	m²/s	square foot per minute	1.55×10^{-3}
Energy				
quantity of heat	kilojoule	kJ	British thermal unit (Btu)	1.055
consumption	gigajoule	GJ	therm	0.1055
consumption	megajoule	MJ	kilowatt hour (kWh)	3.6
Power				
heat flow rate	watt	W	Btu per hour	0.293
motor power	kilowatt	kW	horsepower	0.746
refrigeration	kilowatt	kW	ton	3.52
Heat				
specific heat capacity	kilojoule per kilogram Kelvin	W/kg K	Btu per pound degree Fahrenheit	4.19
specific enthalpy	kilojoule per kilogram	kJ/kg K	Btu per pound	2.33
latent heat	kilojoule per kilogram	kJ/kg	Btu per pound	2.33

[a]Flow rates are sometimes quoted in liters per second (L/s); 1 L/s = 0.001 m³/s.
[b]It is customary to take the unit of atmospheric pressure as the millibar (mbar); 1 bar = 10⁵ Pa. Occasionally, other pressures use this unit.

INDEX

Note: Page numbers followed by *f* indicate figures, *t* indicate tables, and *np* indicate footnotes.

Printed in the United States
By Bookmasters